Issues in Space

The 'Issues in Space' series brings together a broad range of authors and disciplines to promote a diversity of views, perspectives, and critical approaches in the space sector. This series is designed to influence the development of the domestic and global space industry while it is in its formative stage, rather than trying to redress imbalances once they are entrenched in the industry.

This series aims to particularly promote the work of authors from traditionally marginalised groups, including women, early career researchers, and others; voices that are important in discussions of the development of the space industry. The scope of this series covers a broad range of key areas and disciplines that will be explored. Fundamental areas such as: outer space, space law, space policy, space regulation, space archaeology, space ethics, space situational awareness, space traffic management, military activities in space, commercial space, dispute resolution for space activities, space courts, space security, sustained human presence in space, space resource utilisation, space governance, space settlement, armed conflict in space, space resource utilisation, space manufacturing, space tourism, private actors in space, and space mining.

Issues in Space is an interdisciplinary series in that it is devoted to space scholarship but with the added unique focus on the promotion of diversity of authors, disciplines and opinions.

Sandeepa Bhat B. • Adithya Variath
Editors

Artemis Accords and Resource Mining in Outer Space

Springer

Editors
Sandeepa Bhat B.
Centre for Aviation and Space Laws
The West Bengal National University
of Juridical Sciences
Kolkata, India

Adithya Variath
Jindal Global Law School
O.P. Jindal Global University
Sonipat, India

ISSN 2662-902X ISSN 2662-9038 (electronic)
Issues in Space
ISBN 978-981-95-0151-9 ISBN 978-981-95-0152-6 (eBook)
https://doi.org/10.1007/978-981-95-0152-6

This Springer imprint is published by the registered company Springer Nature Singapore Pte Ltd.
The registered company address is: 152 Beach Road, #21-01/04 Gateway East, Singapore 189721, Singapore

Preface

The idea for this book took birth in 2023 during vibrant discussions at the *Global Conference on Resource Mining in Outer Space: Artemis Accords and Beyond* hosted by the Centre of Aviation and Space Laws at the West Bengal National University of Juridical Sciences, Kolkata, the Center for Air and Space Law at the University of Mississippi School of Law, and the Centre for Research in Air and Space Laws at Maharashtra National Law University Mumbai. The event brought together scholars, practitioners, and policymakers from across continents. During the conference, a recurring theme emerged—the urgent need for a comprehensive and critical analysis of resource mining in space, addressing both legal and governance challenges. That moment of intellectual convergence sowed the seeds for this volume. We are deeply grateful to the speakers who shared their perspectives at the conference that laid the foundation for this book. We also thank the contributors who workshopped their research ideas, engaged in debate and deliberations, and refined their arguments throughout the process.

The increasing human and commercial presence in outer space has reignited debates over resource mining beyond Earth. The Artemis Accords, a set of principles guiding space exploration and resource utilisation, have drawn both support and critique from legal scholars, policymakers, and industry stakeholders. This volume is a response to the complex questions surrounding the governance of space resources, sustainability, and international cooperation for protecting the sanctity of the common rights' regime of outer space from the onslaught of commercialisation. This book is the culmination of extensive research, dialogue, and collaboration with experts from diverse backgrounds. Their commitment to advancing discourse on space resource governance has been truly invaluable. This book attempts a comprehensive coverage of each part of the Artemis Accords in light of established space law principles. The uniqueness of this book is the inclusiveness to provide space for varying viewpoints. The contrasting views of contributors are not suppressed, especially to help readers understand the different nuances of space mining and the Artemis Accords. The views expressed are personal to the authors, and the editors of this book may or may not endorse them.

Special thanks are due to Springer for creating a home for this project. We are also grateful to Mamun Mishra and Thirumangai Thamizhmani for their support and belief in this project. This edited volume is the result of a collective effort, and we sincerely appreciate everyone who has contributed to this project in small and big ways. We hope this work contributes to ongoing discussions on space law, Artemis accords, and resource utilisation and serves as a foundation for future research on sustainable governance in outer space.

Kolkata, India
Oxford, UK

Sandeepa Bhat B.
Adithya Variath

Contents

1 Artemis Accords and International Space Law 1
Sandeepa Bhat B.
1.1 Introduction 2
1.2 The Ten Commandments of the Artemis Accords 2
1.3 The Normative Value 5
1.3.1 From a Treaty Perspective 5
1.3.2 Interpretational Significance 6
1.3.3 Customary International Law Perspective 8
1.4 Implications Under the Moon Agreement 9
1.5 The Artemis Accords and the Outer Space Treaty 11
1.5.1 Space Resource Mining 11
1.5.2 Creation of Safety Zones and Preserving the Heritage 12
1.5.3 Peaceful Uses and Debris Mitigation.................. 13
1.6 Concluding Remarks 14
References.. 14

2 The Inevitability of Developments and Changes to Outer Space Property Law Brought About by the Artemis Accords 17
Wian Erlank
2.1 Introduction 18
2.2 The Current Property Law System 19
2.2.1 Background 19
2.2.2 Ownership/Or the Lack Thereof in Terms of the OST 21
2.3 The Artemis Accords 23
2.3.1 Introduction 23
2.3.2 General Commentary on Sections of the Accords........ 24
2.3.3 Space Resources.................................. 26
2.4 Deconfliction of Space Activities: The Safety Zone 29
2.4.1 Possession and Exclusion and Property in Outer Space ... 29
2.4.2 Section 11: Deconfliction of Space Activities with Safety Zones.................................. 31

2.4.3 Allocating Property and Exclusionary Rights: A Possible Solution 34
2.5 Conclusion 35
References 36

3 Mining in the Province of All Mankind: Analysing Legality with Reference to Treaties and Their Interpretation 39
Pranav Satyanath and G. V. Ashok
3.1 Introduction 40
3.2 'Province of All Mankind' and 'Common Heritage of all Mankind': Reviewing the Landscape of International Law 42
3.3 Principles of Interpretation of Treaties and the Outer Space Treaty 46
3.4 Conclusion 49
References 50

4 The Artemis Accords: An Attempt to Internationally Legalise Outer Space Resource Mining and Ownership? 53
Tanushri Joshi
4.1 Introduction 53
4.2 What Are the Artemis Accords? 54
4.3 Why Are the Artemis Accords Stirring Up the Debate? 55
4.4 The Current International Scenario 56
4.5 No Race Between the Space Laws: International Law at Standstill While National Laws Pace Up 58
4.6 Confusion Created by National Laws and International Laws: Is Maritime Law to the Rescue? 60
4.7 Who Prevails Over Whom: National v. International Law? 60
4.8 Why Artemis Accords? What Is the Significance of the Artemis Accords Here? 61
4.9 Lifting the Artemis Accord Veil: Relevance of Soft Law in the Interpretative Process 62
4.9.1 Soft Law as a Step Towards Building Customary International Law 63
4.10 Conclusion 64
References 65

5 The Interplay Between the Moon Agreement and the Artemis Accords: Vienna Convention on the Law of Treaties to the Rescue 69
Atul Alexander
5.1 Introduction 70
5.2 Artemis Accord and the Emerging Principles 71
5.3 Moon Agreement and the Divergence 73
5.4 Moon Agreement and Artemis Accord: The Normative Divide 74
5.5 VCLT and Special Regimes 76

5.6 Artemis Accord and Moon Agreement: Voices from VCLT 76
5.7 Conclusion 82
References... 82

6 **Increasing Fragmentation in Regulating Space Resources: The Imperative of Returning to a Multilateral Approach** 87
Athar Ud Din
6.1 Introduction 88
6.2 Managing Space Resources: Regulatory Gaps and Growing Fragmentation 89
6.2.1 International Law and the Ownership and Use of Space Resources.................................. 89
6.2.2 Growth of National Space Policies 91
6.3 The Artemis Accords: Wither Multilateralism?................ 92
6.4 The Distinctive Position of India.......................... 93
6.4.1 India's Space Policy............................ 94
6.4.2 India and the Artemis Accords....................... 94
6.5 Conclusion 95
References... 96

7 **Artemis Accords as Evolutive Law-Making: Lunar Space Mining and the Rise of Space Militarisation** 99
Adithya Variath
7.1 Introduction 100
7.2 Problematising Artemis Accords and Conceptualising Space Weapons... 102
7.2.1 Inconsistencies of Artemis Accords.................. 102
7.2.2 The Curious Case of Art. IV and Space as an Operational Domain of Warfare.............................. 103
7.3 Artemis Accords, Space Weapons and Lunar Space Mining...... 106
7.3.1 Peaceful Purposes and Space Resources 106
7.4 Artemis Accords as Evolutive Law-Making 109
7.5 Conclusion 111
References... 111

8 **Deconfliction Measures in the Freedom of Use of Outer Space: What Lies Ahead?**.................................... 115
Taufik Rachmat Nugraha and Regi Rivaldi
8.1 Introduction 116
8.2 Deconfliction Principle Under International Space Law 118
8.3 Pseudo-Sovereignty in Space Exploration and Use............. 121
8.4 The Enigma of Space Exploration and Use.................. 122
8.4.1 The Terms of the Freedom of Exploration............. 122
8.4.2 Interpreting the Provisions of the Outer Space Treaty..... 126
8.5 The Rise of Non-state Actors in Space Mining: A New Frontier .. 129
8.6 Conclusion 133
References... 134

9 Artemis Accords and the Developing Countries: A Catalyst for Degenerative Equality and Mutual Benefit of Space/Moon Resources . . . 137
Khafayat Yetunde Olatinwo and Onize Felicity Ifashe
9.1 Introduction . . . 138
9.2 The Early Regimes and Equal Benefit of Space/Moon Resources . 139
9.2.1 Challenges and Criticisms . . . 140
9.3 The U.S. Lunar Program: An Ignored Forewarning . . . 143
9.4 Artemis Accords and Moon Resources . . . 144
9.4.1 Legal Framework for Lunar Resource Utilisation . . . 145
9.4.2 Implications For International Cooperation . . . 145
9.5 Equitable and Beneficial Sharing: The Future of Equal Benefit to Developing Countries in the Artemis Accords . . . 146
9.6 Path Forward . . . 148
9.7 Conclusion . . . 150
References . . . 151

10 India's Tryst with the Goddess of the Moon: The Artemis Accord and Beyond . . . 157
Mohd Sikandar
10.1 Introduction . . . 158
10.2 International Space Regime: No Commercial Licenses, Please . . . 159
10.3 In Unilateral Interpretation, We Believe: The United States Space Programme . . . 161
10.4 Ever evolving and Diversifying: India's Space Programme . . . 164
10.4.1 Militarisation of India's Space Programme . . . 164
10.4.2 Privatisation of India's Space Sector . . . 165
10.5 India and Artemis Accords: A Union with Deep Differences . . . 165
10.5.1 Different Space Objectives . . . 166
10.5.2 Less Developed Private Space Industry . . . 167
10.5.3 Adherence to the Principle of Non-Appropriation . . . 168
10.5.4 No Domestic Space Law . . . 169
10.5.5 Abandoning of the Non-Alignment Doctrine . . . 169
10.6 Conclusion . . . 170
References . . . 170

11 Celestial Property Rights in Outer Space: A TWAIL Exploration . . . 175
Aaditya Vikram Sharma
11.1 Introduction . . . 176
11.2 Framework and Approaches . . . 177
11.2.1 International Space Law . . . 177
11.2.2 The Third World Approaches to International Law . . . 178
11.2.3 Interplay between Third World Approaches and Space Law . . . 179

11.3 Development of Space Law and the Third World 180
11.3.1 The Ad Hoc COPUOS and the Third World 181
11.4 The Outer Space Treaty 182
11.5 The Moon Agreement and Third World Countries 186
11.5.1 Genesis of the Moon Agreement: A Brief Synopsis 186
11.5.2 Drafting of the Moon Agreement and Third World Countries 187
11.6 Conclusion 195
References.. 197

Index 203

About the Contributors

Atul Alexander is an Assistant Professor at the West Bengal National University of Juridical Sciences, Kolkata, India. He was short-listed as a visiting research scholar at the Stockholm Center for International Law, US Naval War College, United States. He completed his Diploma in 'Law of the Seas' magna cum laude from the Korean Maritime Institute and Siracusa International Institute for Criminal Justice and Human Rights, Italy, with scholarships in 2022 and 2023, respectively. He has contributed to leading journals published by Cambridge University Press, Springer, Brill, Oxford University Press, etc. He is also a Member of the Australian and New Zealand Society of International Law.

G. V. Ashok is a Partner at Factum Law and Allied Law Practices and an Advocate practising before the High Court of Karnataka in Bengaluru, India. He advises the technology sector on intellectual property rights, contracts, risk management, and technology transfers, addressing both domestic and cross-border legal requirements. In space law, he advises governments and private entities on regulations governing commercial space activities, space situational awareness, international trade and investment law, and public international law. He is frequently invited by governments, think tanks, academic institutions, and publications in India and abroad to lecture and publish on space law, policy, and cybersecurity.

Sandeepa Bhat B. is working as a Professor of Law and the Director of the Centre for Aviation and Space Laws at the National University of Juridical Sciences, Kolkata. He holds the unique feat of being the first Indian to be inducted as the UNIDROIT Correspondent for India. He has been a member of four distinguished international bodies, including the American Society of International Law. Dr. Bhat has published six books on space law, four books on medical law, and two more on aviation law. In addition, he has published more than seventy articles in the journals of international and national repute. He is the recipient of the inaugural Sachdeva Award for Space Law 2022 in recognition of his outstanding contribution to space law. He has presented over two hundred research papers at international and national conferences, including the coveted International Astronautical Congress. He also has the distinction of being a member of the Indian Space Research Organization's Expert Group for drafting the National Space Act for India.

Wian Erlank works as a Professor of Law at North-West University, South Africa, where he currently teaches Space Law and Cyber Law. He is an Advocate of the High Court of South Africa, a Fellow of the European Law Institute and a Member of the International Institute of Space Law. He serves as Editor-in-Chief of the *Potchefstroom Electronic Law Journal (PER/PELJ)* as well as on the editorial board of Lex ad Coelum. His research in the field of space law is focused on property rights in outer space.

Onize Felicity Ifashe is pursuing her Barrister-at-Law program at the Nigerian Law School, Yola, Felicity. She is an alumnus of the Kwara State University, Malete, Ilorin, Nigeria, graduating with a First Class Honours in Law and earned a Distinction in Common Law from Kwara State College of Arabic and Islamic Legal Studies and was recognised as the Best Graduating Student in the Faculty of Law Kwara State University, Malete (2022).

Tanushri Joshi is a space and technology lawyer currently working as a Senior Legal and Business Development Manager at HEX20. Previously, she worked with Skyroot Aerospace, where she was responsible for legal and business development functions, she also drafted India's first private launch services agreement and was also a part of India's first private rocket launch, Vikram-S. She has extensive experience in the space sector ranging from launch vehicles to satellites, advising on legal matters and business strategies to overcome regulatory challenges, formulating and fostering strategic partnerships globally, generating sales and contributing to business expansion as well. She is a member of the International Institute of Space Law's (IISL) Working Group on Light Pollution of the Night Sky from a Space Law Perspective.

Taufik R. Nugraha is the Director and Research Fellow at the Centre for Air and Space Policy (CASP) in Indonesia. He also works as a visiting lecturer at several Indonesian universities, sharing his expertise in space law, policy, and security. He is also a Space Law and Policy Group member at the Space Generation Advisory Council. Moreover, Taufik is a research fellow at the Indonesian Centre for Law of the Sea (ICLOS) at Universitas Padjadjaran and has completed an ASEAN research fellowship at the Centre for International Law at the National University of Singapore. His research focuses on Space Law and Policy, Astropolitics, Space Security, Ocean Law and Policy, and Nuclear Law.

Khafayat Yetunde Olatinwo is an Associate Professor of International Space Law in the Department of Jurisprudence & Public Law, Faculty of Law, Kwara State University, Malete. She has written over 30 articles in referred local, national, and international journals and is a member of academic and professional bodies within and outside Nigeria. She is currently the Vice-president of the Space Law and Arbitration Association, Nigeria and the Head of Department, Department of Jurisprudence & Public Law, Faculty of Law, Kwara State University.

Regi Rivaldi is a staff member at the Internationalisation Office and a legal researcher at Universitas Padjadjaran. He also works as a legal auditor and lawyer at two law firms in Indonesia. His research primarily focuses on space law, technology law, public international law, and investment law.

Pranav Satyanath currently leads the policy affairs and government relations efforts at *SkyServe*, an AI & edge computing technology company for satellite platforms. Pranav's research interests lie at the intersection of technology and policy, and he has written reports, essays, and op-eds on topics such as risk reduction in outer space, international space cooperation, and the politics of nuclear weapons. His other academic interests include the history of arms control negotiations and the social shaping of technology.

Aaditya Vikram Sharma is an Assistant Professor at the Vivekananda School of Law and Legal Studies, Vivekananda Institute of Professional Studies - Technical Campus, New Delhi. He is currently pursuing his PhD in Space Law at the West Bengal National University of Juridical Sciences (WBNUJS), Kolkata. Aaditya's research interests are centred around Space Law, reflecting his passion for exploring legal frameworks governing outer space. He recently participated in the Southern Hemisphere Space Studies Program at the International Space University, France. Aaditya also holds an Executive Certificate in Space Studies from the University of South Australia, Adelaide.

Mohd Sikandar is an Assistant Professor of Law at National Law University Odisha. In particular, he specialises in international law and related subjects. After completing his master's from NALSAR University of Law, he has written extensively on topics of national and international importance ranging from the rights of the Uyghur community to the global rise of populism and to the most recent Israel versus South Africa case at the ICJ. The emergence of novel technologies like AI and their impact on human rights is another field in which he holds an interest.

Athar Ud Din is an Assistant Professor and In-Charge Head of the Center for International Relations at the Islamic University of Science and Technology. He holds a Ph.D. in International Legal Studies from Jawaharlal Nehru University, India. His academic interests relate to various aspects of international relations and law, especially developing countries in international relations, management of the Global Commons, and the principle of Common but Differentiated Responsibilities in environmental treaties.

Adithya Variath is an Assistant Professor at Jindal Global Law School, O.P. Jindal Global University, India. He pursued an MSc in Global Governance and Diplomacy at the Oxford Department of International Development, University of Oxford as an Oxford India Centre for Sustainable Development's Elizabeth Moir Scholar. Before moving to Oxford, he was an Assistant Professor of Law and the founding Coordinator of the Centre for Research in Air and Space Laws at Maharashtra

National Law University Mumbai (MNLU-M), India. He was also the Academic Supervisor of the TradeLab Law Clinic and a Member of the Research Council at MNLU-M. He also serves as the Editor of the *Indian Review of Air and Space Law*. His research focuses on space law, aviation law, emerging technologies, and global institutional design.

Chapter 1
Artemis Accords and International Space Law

Sandeepa Bhat B.

Abstract The twenty-first century accomplishments in the sphere of space activities have gone much beyond past human imaginations. After the dream to enter outer space and to land on the celestial bodies has become reality, the next level of thinking in terms of exploitation of celestial resources has started. It is not only States but also private entities that have started to enter the realm of outer space for various commercial activities, and the latter are not far behind even in their plans to exploit celestial resources. In this era of a race to celestial resources, the Artemis Accords of the United States made headlines in 2020. While showcasing it as the sustainable practice of space resource exploitation, the United States has sought international cooperation by acceding to the Artemis Accords. By highlighting the non-binding nature of the Artemis Accords, the United States has asserted that they are not in violation of any international space treaty. The number of States joining the Artemis Accords is increasing regularly. In this context, it has become significant to evaluate the status of the Artemis Accords under the gamut of international space law. Though the Moon Agreement 1979 is explicitly against such select-few grabbing celestial resources, its limited ratification by States makes it ineffective. Hence, the crucial debate is on the compatibility of the Artemis Accords with the Outer Space Treaty 1967, which provides for fundamental principles of space law.

Keywords Artemis Accords · Commercialisation · Outer Space Treaty · Moon Agreement · Resource mining · VCLT

S. Bhat B. (✉)
Centre for Aviation and Space Laws, The West Bengal National University of Juridical Sciences, Kolkata, India
e-mail: sandeep@nujs.edu

S. Bhat B., A. Variath (eds.), *Artemis Accords and Resource Mining in Outer Space*, Issues in Space, https://doi.org/10.1007/978-981-95-0152-6_1

1.1 Introduction

Space activities have shifted their dimensions in the recent past. While the twentieth century marked the predominance of scientific exploration in space, the twenty-first century is evidencing wide-scale commercial space activities. Telecommunication, Internet, direct television broadcasting, remote sensing, etc., are all classic examples of the use of outer space for commercial activities since a reasonably long period of time. However, these activities involve only the use of outer space without extracting any resources from there. In contrast, the recent developments in some States indicate their zeal for extracting resources from outer space and celestial bodies.

The first major development in terms of granting legal recognition for celestial resource exploitation took place in the United States with the passing of the Commercial Space Launch Competitiveness Act 2015. This has been followed by Luxembourg (Law of July 20, 2017), the United Arab Emirates (*Law on the Regulation of the Space Sector,* 2019) and Japan (*Act on the Promotion of Business Activities for the Exploration and Development of Space Resources,* 2021) to confer similar rights to exploit celestial resources through the legislative process. India followed this pursuit with an executive process of passing the Indian Space Policy 2023. Meanwhile, the National Aeronautics and Space Administration (NASA) promulgated the Artemis Accords as an international cooperative arrangement in 2020. Titled "Principles for Cooperation in the Civil Exploration and Use of the Moon, Mars, Comets, and Asteroids for Peaceful Purposes" and described as nonbinding political commitments, the Artemis Accords is directed towards commercial exploitation of celestial resources. Since its inception, the Artemis Accords has triggered debates in terms of its compatibility with obligations under the United Nations space treaties. While the current number of parties to it is increasing, standing at 53 as of 21 January 2025 (NASA, 2020), the legality of the Artemis Accords cannot be judged on this factor. Hence, evaluating the Artemis Accords in light of obligations under international space treaties is of utmost significance.

1.2 The Ten Commandments of the Artemis Accords

The Artemis Accords begins with underscoring the significance of space exploration, use and cooperation in yielding benefit to all humankind. It stipulates that we are in the new age of exploration, and hence, going beyond the Apollo programme is essential. In this context, it espouses the Artemis Program for the sustainable human exploration of the solar system. This programme is depicted to be based on the cooperation with international and commercial partners of NASA. Except the Moon Agreement 1979 (United Nations, 1984), the opening part of the Artemis Accords recognises the need for adhering to obligations under the remaining four space treaties entered under the aegis of the United Nations (United Nations, 1967, 1968, 1972, 1975).

Advancing the purpose of the Artemis Program as civil exploration and use of outer space, the Artemis Accords is set to establish a common vision based on a set of principles and guidelines that are practical in nature. Adherence to these principles is expected to help in improving safety, mitigating uncertainties, and promoting sustainable and beneficial use of outer space. Significant obligations of the Outer Space Treaty and other instruments are also said to be operationally implemented by the Artemis Accords, in Section 1 (NASA, 2020). Section 2 of the Artemis Accords calls for implementing it in all cooperative activities by inserting appropriate provisions in the cooperation agreement/instrument signed by the parties. Hence, the Artemis Accords is aimed at developing bilateral or plurilateral cooperation in the exploration and use of outer space.

The next ten provisions of the Artemis Accords stipulate the ten cardinal principles to be followed by the parties. The principle of peaceful uses is the first among them. The cooperative activities under Section 3 of the Artemis Accords are explicitly stipulated to be for peaceful purposes exclusively by adhering to applicable international law. Section 4 of the Artemis Accords is on maintaining transparency on two counts. One of the aspects is the dissemination of information about national space policies and plans of space exploration by States on the basis of rules and regulations applicable at their domestic levels. The other aspect emphasises the need for carrying forward the objective of Article XI of the Outer Space Treaty by sharing the scientific information emanating from space activities under the Artemis Accords.

Interoperability is the next principle enshrined under Section 5 of the Artemis Accords. The need for developing interoperable and common exploration infrastructure and standards is recognised here for enhancing the outcomes of space activities. This has been portrayed as the spirit of cooperation between participating States. A reference to the "commercial utilisation" as an outcome of space activities appears for the first time in this section of the Artemis Accords. Carrying forward the spirit of cooperation, Section 6 envisages the fulfilment of obligations under the Rescue Agreement by providing emergency assistance to personnel in outer space who are in distress.

Section 7 insists on the compliance of the obligation to register space objects under the Registration Convention. Since the Artemis Accords advocates for joint ventures and there cannot be more than one registering State for a space object as per the Registration Convention, the participating States must collectively determine which one of them shall register the space object. The next provision, Section 8, is on the release of scientific data obtained from cooperative activities under the Artemis Accords. Though the general commitment to transparency is reiterated here, the provision stipulates two exceptions to it. First, the signatories can release only information regarding their own activities. Other participating States' scientific information can be released only with necessary coordination with them and subject to the protection of their rights, proprietary or otherwise. Secondly, States are not entitled to share the scientific data resulting from private sector operations.

Preserving outer space heritage is the next principle enshrined under Section 9 of the Artemis Accords. Describing outer space as our shared heritage, it calls for developing mutually agreed standards and practices for preserving historic landing sites, artefacts, spacecraft and other evidences of human activities on celestial bodies. Contributing to future multilateral efforts in this regard is also mentioned under the provision.

The most controversial parts of the Artemis Accords are Sections 10 and 11. The former provision deals with space resources and recognises that the utilisation of space resources is beneficial to humankind. The extraction and utilisation of space resources are said to be conducted in compliance with the Outer Space Treaty. In the zeal to justify resource exploitation, it is also declared in Section 10(2) that the extraction of space resources does not amount to national appropriation under Article II of the Outer Space Treaty. This is further coloured with the commitment of States to maintain transparency by openly disclosing their activities relating to space resource extraction. The experience of States under the Artemis Accords, as reflected in Section 10(4), is also enshrined as a pathbreaker in developing a multilateral regime for the extraction and utilisation of space resources.

Section 11 deals with the deconfliction of space activities to ensure safety and to prevent harmful interference with each other's space activities. It centers around the principles of due regard and consultation, which are stipulated under Article IX of the Outer Space Treaty. Due consideration of the United Nations Guidelines for the Long-term Sustainability of Outer Space Activities, which was adopted in 2019 by the UNCOPUOS, is also stipulated in this provision. While emphasising the prevention of harmful interference with others' activities, furnishing of necessary information about the location and nature of space activities and establishment of safety zones are highlighted as necessary prerequisites. The subsequent clauses of this section delve into the details of safety zones. They stipulate the need for considering the nature of operations, the environment of operations and reasonable parameters in determining the size and scope of safety zones. The nature and existence, as well as the size and scope of safety zones, may change with any change in activities. The safety zones are mentioned to be temporary and cannot stretch beyond the time of concerned operations for which safety zones are established. The establishment, alteration or ending of safety zones shall be notified to the signatory States and to the United Nations Secretar- General.

Section 11 also stipulates the application of national rules and regulations of the State concerned in providing information to a party to the Artemis Accords regarding the basis for an area declared as a safety zone. Subject to proprietary and export control requirements as well as to the extent practicable and feasible, transparency needs to be maintained by States regarding the general nature of operations taking place in their safety zones. Prior notification and coordination among the parties to the Artemis Accords is stressed upon to prevent harmful interference with each other's activities in safety zones. It is enshrined in Section 11(10) of the Accords. While highlighting the flexibility of safety zones, encouraging scientific discovery, technology demonstration, and extraction and utilisation of space resources are emphasised. The commitment to allow free access to all parts of celestial bodies and adherence to other provisions of the Outer Space Treaty are reiterated under Section

11. Based on mutual experiences and consultations with the international community, necessary adjustments in the use of safety zones are also contemplated.

The last principle under the Ten Commandments deals with orbital debris, which is mentioned in Section 12. The signatories to the Artemis Accords are obligated to have a plan for end-of-mission debris mitigation, including the passivation and disposal of space objects in a safe and efficient manner. In case of joint missions, one of the States needs to be identified as primarily responsible for taking necessary steps in this direction. Ensuring safety in flight profiles and operational configurations and proper disposal of space structures after missions are stressed upon to limit debris creation by various ways.

The final provision of the Artemis Accords (Section 13) stresses on the need for periodic consultation to review the implementation of the above principles as well as to explore the areas of future cooperation via the exchange of views. While the eligibility for the registration of this document under Article 102 of the Charter of the United Nations is negated, it is unilaterally declared as a document to be circulated among the members of the United Nations as an official document of the United Nations.

1.3 The Normative Value

The binding nature of international norms emanates from treaties and customary international law (Jia, 2010). Hence, in evaluating the normative value of the Artemis Accords, we need to look at it from the perspectives of treaty-making and interpretation as well as the development of customary international law.

1.3.1 From a Treaty Perspective

The Vienna Convention on Law of Treaties 1969 (hereafter 'VCLT') is the key document in understanding the definition of a treaty and the treaty interpretation (United Nations, 1969). "Treaty", as per Article 2(1)(a) of the VCLT, means "an international agreement concluded between States in written form and governed by international law, whether embodied in a single instrument or in two or more related instruments and whatever its particular designation." The drafting history of the VCLT reflects the understanding of the negotiators that the words "governed by international law" includes the intention of States to establish a legal relationship or to establish rights and obligations between them, which is the element of consensus *ad idem* as understood under the traditional contract law (Mardani & Hooshmand, 2016). Such an intent of parties can be derived from the language used in the treaty, negotiating history or the registration of the treaty under Article 102 of the United Nations Charter. The Artemis Accords is explicitly mentioned to be a mere nonbinding political commitment, which is not eligible for registration under Article 102 of the United Nations Charter. Even the language used in the Accords, like 'recognize', 'commit', 'intend', 'note', 'emphasize', 'plan' etc., is far from creating any binding obligation on States. Hence, the Artemis Accords can never qualify the requirement of establishing a legal relationship based on the rights and obligations of States.

1.3.2 Interpretational Significance

Some writers have mooted the possibility of using the Artemis Accords as an interpretative tool under the VCLT. This line of argument is more relevant from the perspective of sections 10 and 11 of the Artemis Accords, which are controversial from the standpoint of the Outer Space Treaty. Under public international law, Boyle has argued the role of soft law in treaty interpretation by reflecting the subsequent State practice in the application of the treaty as a tool of interpretation under Article 31(3)(b) of the VCLT (Boyle, 2018). In its Seventieth Session, the International Law Commission (ILC) has observed that if the subsequent State practice falls under Article 31(3)(b) of the VCLT, it stands as an authentic means of treaty interpretation (ILC, 2018). It is considered to be an objective evidence of parties' understanding of treaty provisions. In the same Session, the ILC has provided the following definition in Conclusion 4:

> A subsequent practice as an authentic means of interpretation under article 31, paragraph 3 (b), consists of conduct in the application of a treaty, after its conclusion, which establishes the agreement of the parties regarding the interpretation of the treaty.

By referring to Boyle's writings and ILC conclusions, Neef argues that the Artemis Accords will influence the interpretation of the Outer Space Treaty provisions especially in light of existing confusion on what is permitted and what is not permitted under the Outer Space Treaty (Rachel, 2021). It is also argued to be an instrument to fill gaps in outer space law and governance as a strong tool of interpretation under Article 31(3)(b) of the VCLT.

On a contrary note, Deplano describes the Artemis Accords as an instrument of one State, which cannot qualify as a reference under Article 31(3)(b) of the VCLT (Deplano, 2021). International dispute settlement bodies like the Appellate Body of the World Trade Organisation (WTO) and the Tribunals of International Centre for Settlement of Investment Disputes (ICSID) have rejected to use such individual views of States in interpreting treaties.[1] The WTO Appellate Body and Panels have consistently negated the practice of zeroing employed by select-few States,

[1] Note: References are made to Japan: Alcoholic Beverages II, Report of the Appellate Body WT/DS8/AB/R, WT/DS10/AB/R and WT/DS11/AB/R, 1 November 1996; Sempra Energy International v Argentine Republic, ICSID Case No ARB/02/16, 28 September 2007; and Philip Morris Brands SÀRL, Philip Morris Products SA and Abal Hermanos SA v Oriental Republic of Uruguay, ICSID Case No ARB/10/7, 28 June 2016.

including the United States and the European Union, in the determination of dumping under the Anti-dumping Agreement.[2] This approach is also adopted by the International Court of Justice (ICJ) in the *Whaling in the Antarctic* case to reject the use of the International Whaling Commission resolution in interpreting the provisions of the International Convention for the Regulation of Whaling 1946 (International Court of Justice, 2014). In summarising the ICJ's verdict, Arato mentions the following:

> ...while unanimous or consensus resolutions of a supervisory treaty body might be considered subsequent agreements or practice relevant to the interpretation of the underlying convention, resolutions adopted by disputed majority will not count under the general rule of interpretation (Arato, 2014).

In the Seventieth Session, the ILC had also clarified in Conclusion 10 that an agreement under Article 31(3)(b) of the VCLT "requires a common understanding regarding the interpretation of a treaty which the parties are aware of and accept" (ILC, 2018). Since no such common understanding of either the Outer Space Treaty or other United Nations space treaties is reflected in the Artemis Accords, it fails to qualify the pre-requisites of State practice under Article 31(3)(b) of the VCLT.

Deplano does not rule out the relevance of the Artemis Accords to conceptualise as a subsequent practice under Article 32 of the VCLT (Deplano, 2021). As per ILC Conclusion 4(3), "A subsequent practice as a supplementary means of interpretation under article 32 consists of conduct by one or more parties in the application of the treaty, after its conclusion". However, the role of subsequent practice as a supplementary means of interpretation shall not be overemphasised for two main reasons. First, as mentioned under ILC Conclusion 7(3), the parties to a treaty should not use a practice to amend or modify a treaty obligation as the amendment or modification of a treaty by a subsequent practice has not yet gained the general acceptance. This is also reflected in the rejection of States at the Vienna Conference of 1968 to adopt Article 38 of the Draft Articles on the Law of Treaties, which was intended to allow modification of a treaty by subsequent practice (United Nations, 1968). Even the United States representative at the Vienna Conference argued for deleting the provision by expressing the following concern:

> What particularly worried the United States delegation was that relatively low-ranking officials such as vice-consuls and third secretaries might interpret a treaty erroneously and follow a course of conduct which, unknown to governments, could lead to modification of the treaty (United Nations, 1968).

Moreover, the 'practice' under Article 32, being much inferior to that of practice under Article 31(3)(b), can never override or rewrite treaty obligations in favour of

[2] Note: European Communities—Anti-Dumping Duties on Imports of Cotton-type Bed Linen from India, Report of the Appellate Body WT/DS141/AB/R, 1 March 2001; United States—Final Dumping Determination on Softwood Lumber from Canada, Report of the Appellate Body WT/DS264/AB/R, 11 August 2004; United States—Measures Relating to Zeroing and Sunset Reviews, Report of the Appellate Body WT/DS322/AB/R, 9 January 2007; United States—Use of Zeroing in Anti-Dumping Measures Involving Products from Korea, Report of the Panel WT/DS402/R, 18 January 2011.

or against any State. Secondly, any process of treaty interpretation shall not take away the stability and predictability of the treaty.[3] Individualistic or select-few States' approaches carry the risk of taking away such stability and predictability, acceptance of which might ultimately result in the redundancy of the treaty. Although, ILC has recognised the usefulness of subsequent practice in "finding a flexible approach to treaty application and interpretation", it has explicitly stipulated the requirement of it being rational and predictable (ILC, 2008). Hence, the need for providing an ordinary meaning, in the context and in the light of the object and purpose of the treaty cannot be undone by any practice to take away the treaty's stability (Bianchet, 2017).

1.3.3 Customary International Law Perspective

The argument that the Artemis Accords may contribute to the emergence of customary international law is advocated by some writers. Smith, for example, advocates for a US-centric good governance model to be developed by transforming the Artemis Accords into customary international law (Smith, 2021). Fidler opines that the principles mentioned in the Artemis Accords "can generate state practice and create precedents" (Fidler, 2020). These arguments seem to be fascinating and reflecting the flexibility needed for furthering the commercial interest in space. However, they completely disregard the history and foundations of international space law. If individual interest, commercial or otherwise, was paramount; there was no requirement of regulating the outer space with an international regime. The self-centred and hegemonic domestic laws could have been made applicable for outer space activities based on colonial experiences. This approach was not followed to ensure that the colonial conflicts and tragedies would not resurface in the space era. The States' representatives at the United Nations, negotiators of space treaties and early scholars of space law were clear on this point, and accordingly, they resorted to the application of a common rights' regime to outer space and celestial bodies (Christol, 1998). The Outer Space Treaty is a living example of this common *opinio juris* reflected widely in its provisions.

The customary international law presupposes the presence of *opinio juris* and State practice. Cheng puts forth his idea of instant customary international law and argues that the *opinio juris* is the single constituent element of customary international law (Cheng, 1997). The *opinio juris*, being a subjective element, is objectively assessed, usually with the help of State practice. Thus, the latter plays only a

[3] Note: A reference can be made to Article 3.2 of the WTO Dispute Settlement Understanding, which explicitly stipulates the need for protecting the security and predictability of the multilateral trading system in the process of dispute settlement. This has been religiously followed by the WTO dispute settlement bodies. See, for example, European Communities—Customs Classification of Certain Computer Equipment, Report of the Appellate Body WT/DS62/AB/R, WT/DS67/AB/R and WT/DS68/AB/R, 5 June 1998.

supplementary role, possibly as an evidence regarding the presence of *opinio juris* (Guzman, 2008). This means that a mere presence of a State practice, howsoever long it might be, will not create customary international law. What is required is the adequate reflection of the positive *opinio juris*. ICJ's judgement in the *North Sea Continental Shelf Cases* reflects this in the following words:

> Not only must the acts concerned amount to a settled practice, but they must also be such, or be carried out in such a way, as to be evidence of a belief that this practice is rendered obligatory by the existence of a rule of law requiring it.... The States concerned must therefore feel that they are conforming to what amounts to a legal obligation (*North Sea Continental Shelf Cases,* 1969, *p. 44, para. 77*).

The Artemis Accords is a mere nonbinding political commitment, and the States are signing it with a clear idea of it being not a legal obligation. Any practice developed under it will remain as self-centred and exclusionary practices of a few States, falling short of reflecting common *opinio juris*. The object and purpose of the Outer Space Treaty, with a strong inclination towards common rights' regime, reflect the common *opinio juris* of States at a multilateral level, which cannot be negated or altered with a nonbinding plurilateral Artemis Accords or practices emanating from it. While a State that has signed and not ratified a treaty is obligated to respect the object and purpose of the treaty as per Article 18 of the VCLT, the States that have already ratified have much higher obligations. Developing any practice contrary to a multilateral treaty obligation is just a reflection of convenient interpretation rather than the existence of a positive *opinio juris* to formulate customary international law. Accepting such practices as overriding custom is also against the United States' own stand in the Vienna Conference 1968 while rejecting the adoption ofArticle 38 of the Draft Articles on the Law of Treaties.

Russia and China, two other major spacefaring States, have opposed the Artemis Accords in clear terms. While criticising the US-centric approach, Russia was vocal against the United States' efforts to privatise and colonise space (Goswami, 2021). It has viewed the Artemis Accords as a departure from cooperation and mutual support that has been developed by them under the International Space Station project (Cowing, 2020). China equated the United States' efforts to the land-grabbing methods employed by colonial powers in history (Ji et al., 2020). As opposed to the Artemis Accords, Russia and China have jointly sought cooperation for their International Lunar Research Station (ILRS) project. With this, the conflicts that were avoided in the 1960s and 70 s by accepting a common rights' regime under the United Nations space treaties are on the brink of resurfacing in the contemporary world.

1.4 Implications Under the Moon Agreement

The Moon Agreement incorporates the most stringent form of the common rights' principle, which is the common heritage of mankind (CHM). Article 11 of the Moon Agreement details in its eight paragraphs the application of CHM to the celestial bodies and their natural resources (Bhat, 2007). While the freedom of exploration and use of celestial bodies is kept intact (Moon Agreement, Art. 11(4)), the exploitation of natural resources of celestial bodies is subject to the establishment of an international regime (Moon Agreement, Art. 11(5)). Such an international regime is to be developed by States Parties to the Moon Agreement in accordance with Article 18 of the Moon Agreement. The Artemis Accords does not qualify as an international regime under the Moon Agreement. Moreover, the Artemis Accords fail to incorporate the most significant element of CHM in the extraction and utilisation of space resources, which is the equitable sharing of benefits (Moon Agreement, Art. 11(7)). The Hague International Space Resources Governance Working Group (The Hague International Space Resources Governance Working Group, n.d.), which parallelly developed building blocks on space resource activities, has also recognised the mandate of benefit sharing.[4] Hence, justifying the Artemis Accords in the context of the Moon Agreement is beyond the realm of law. This is also reflected in the silence of the Artemis Accords regarding the compliance with obligations under the Moon Agreement, despite affirming the importance of compliance with the other four United Nations space treaties.

After the withdrawal of Saudi Arabia, the Moon Agreement has 17 parties, and 4 signatories who have not ratified the Moon Agreement. Saudi Arabia's withdrawal from the Moon Agreement was with a clear intent to join the Artemis Accords. Among the remaining, Armenia, Australia, Austria, Belgium, Chile, Mexico, Netherlands, Peru and Uruguay are parties to both the Moon Agreement and the Artemis Accords. France, India and Romania are part of the Artemis Accords and signatories of the Moon Agreement without ratification. Hence, this brings forward an interesting debate over the status of three different categories of States. The first category includes obligations of parties to the Moon Agreement on the basis of *pacta sunt servanda*. Secondly, the obligations of signatories to the Moon Agreement, who have not ratified it, in the context of Article 18 of the VCLT. Thirdly, the possibility of application of customary international law principles to States that are not parties to the Moon Agreement. The status of Saudi Arabia may also qualify as distinct among the third category of States.

Having understood the basis of debate under the Moon Agreement, we must acknowledge the fact that this issue is substantially an academic debate. While the Moon Agreement has been ratified by a limited number of States, none of them has activities on celestial bodies. In addition, the Moon Agreement in Article 17 explicitly allows amendments to it, which can be brought into force with the acceptance by a majority of parties to the Moon Agreement. Hence, the above-mentioned nine parties to the Moon Agreement, who are also part of the Artemis Accords, may use Article 17 of the Moon Agreement to suit their way. Since there is a separate chapter on the Moon Agreement as a part of this book, the present chapter will not delve further into the interplay between the Artemis Accords and the Moon Agreement.

[4] Note: Building Blocks for the Development of an International Framework on Space Resource Activities, 12 November 2019, Clause 13. Reference made here is only to highlight the significance of benefit sharing. The method of benefit sharing shall be negotiated multilaterally by keeping in mind the interests of all stakeholders.

1.5 The Artemis Accords and the Outer Space Treaty

The Outer Space Treaty is the Magna Carta of international space law (Andem, 2004; Kopal, 1998, p. 116). It provides several fundamental principles of space governance, which have attained the status of universally binding customary international law (Qizhi, 1997). Derogation from these principles through the Artemis Accords or practices under it will be a violation of international law. A mere assertion of compliance with the Outer Space Treaty under the Artemis Accords is inadequate to justify any action which is in contravention of the Outer Space Treaty. Hence, probing into the areas of overlap between the Artemis Accords and the Outer Space Treaty is of utmost importance.

1.5.1 Space Resource Mining

Article II of the Outer Space Treaty contemplates non-appropriation as a fundamental principle of space law. It reads as follows:

> Outer space, including the moon and other celestial bodies, is not subject to national appropriation by claim of sovereignty, by means of use or occupation, or by any other means.

While allowing the mining of space resources, Section 10 of the Artemis Accords declares that "extraction of space resources does not inherently constitute national appropriation under Article II of the Outer Space Treaty." It is interesting to note here that such kind of assertions is not found in any other document of international law. The drafters' scepticism regarding compliance with Article II of the Outer Space Treaty is clearly evident in this provision. Moreover, the question of Section 10 of the Artemis Accords complying with the principle of non-appropriation under the Outer Space Treaty is a matter of legal determination and not a part of political assertion.

From a legal perspective, the principle of non-appropriation under Article II of the Outer Space Treaty is the custodian of the common rights' regime developed in space law. The space resources, being bounties of nature, belong to humankind as a whole (Rao, 1997). Extraction and use of them by a select-few, individuals or States, is against the foundations of the common rights' regime. Thus, the term "national appropriation" in the provision is not confined to States, and Article II's prohibition is equally applicable to international organizations as well as juridical and natural persons (Christol, 1984). It is illogical to give a narrow meaning to "national appropriation" by arguing that only State is prohibited under Article II but not an individual or a group of States. This fact is also clear from the use of the term "national activities" under Article VI of the Outer Space Treaty, which covers the activities of both non-governmental entities and international organisations along with that of States.

The argument that Article II's prohibition is confined to land territories of celestial bodies and does not extend to their natural resources has no merit. The non-appropriation principle is not just confined to the means like claim of sovereignty, use or occupation, but extends to "any other means". The use of "any other means" is wide enough to go beyond territorial rights to cover the extraction and use of space resources. Brooks has rightly remarked:

> Had the Space Treaty drafters intended only a territorial ban they would not have included 'means of use' or 'any other means' in amplifying the ban on national appropriation, since… those terms do in fact move beyond mere territorial appropriation (Brooks, 1969, p. 342).

Allowing the extraction and use of space resources by a select few States goes against the principle of equity, which is strongly embedded in the Outer Space Treaty. The mandate of carrying on outer space activities "for the benefit and in the interest of all countries" and adhering to the principle of the province of all [hu]mankind under Article I of the Outer Space Treaty needs a special mention here. It is not just a ritualistic provision to please the developing States but a legal commitment undertaken by all States. This is adequately reflected by the use of the term 'shall' in the provision. Without respecting equity, the interests of developing States will never be served (Steffen, 2022). Hence, any regime of space mining that disregards the benefit and interest of other States is also afoul of Article I of the Outer Space Treaty.

1.5.2 Creation of Safety Zones and Preserving the Heritage

Rosy justifications are provided under Section 11 of the Artemis for the creation of safety zones by the parties to the Artemis Accords. The need for preventing harmful interference with other States' activities under Article IX of the Outer Space Treaty is highlighted as the basis for creating safety zones. While Article IX of the Outer Space Treaty only underscores the need for international consultation to prevent harmful interference, it does not permit the creation of safety zones by States. Any unilateral declaration of safety zone by States will be in violation of Articles I and II of the Outer Space Treaty (Nelson, 2020).

The freedom of exploration, use and scientific investigation is ensured to all States without any discrimination under Article I of the Outer Space Treaty. Allowing free access to all areas of celestial bodies without asserting individual rights is also a commitment undertaken by the parties to the Outer Space Treaty. The State-declared safety zones under the Artemis Accords take away the freedom and free access guaranteed to all other States. Such safety zones also carry the elements of sovereign rights to outer space and celestial bodies. These elements of sovereign rights have the potentiality to exclude others' access to safety zones under the pretence of essential safety measures. This will be in violation of the non-appropriation principle under Article II of the Outer Space Treaty. The safety zones favour the first-mover advantage and breed hostilities among competitors. Hence, the creation of any safety zone needs to be carefully negotiated with the consent of all stakeholders (Schingler, 2020).

Though comparatively a lesser evil, the provision on outer space heritage preservation is not free from concerns. The determination of outer space heritage is subjective in nature. The vague narration of "historically significant human or robotic landing sites, artefacts, spacecraft, and other evidence of activity on celestial bodies" as part of outer space heritage under Section 9 of the Artemis Accords opens Pandora's box for future interpretation. It may end up as an unbridled discretion to a State or a few to define their outer space heritage in their own way, much to the detriment of others. Undefined measures for the preservation of outer space heritage may take different shapes in future; which may include preventing entry, constructing protective installations, implementing safety or security protocols, etc. These measures, if taken, will violate Articles I and II of the Outer Space Treaty.

1.5.3 Peaceful Uses and Debris Mitigation

The Artemis Accords contains clauses on peaceful uses and debris mitigation plans. Though they seem to further the cause of the Outer Space Treaty, vagueness and loopholes persist. The expression "[activities] exclusively for peaceful purposes" retains the historic interpretational debate between "non-military" and "non-aggressive" activities (Cheng, 2000). The contemporary world has moved in favour of "non-aggressive" activities, either by express declarations or by implied practice, which further complicates the affairs. What activities are permissible under the non-aggressive category and what are prohibited by being aggressive is again a subjective element to be determined. Since the Artemis Accords promotes cooperative activities between partnering States, permitting non-aggressive military uses will increase distrust and hostilities among different block.

Finally, Section 12, the provision on debris mitigation plans, is vague and has a limited scope. The provision addresses only mitigation of debris and leaves out other forms of environmental damage that may be caused by chemical, biological or radiological pollutants (Bhat, 2014). Moreover, the commitment of States is confined to outer space since only orbital debris is part of mitigation plans. This leaves out the entire gamut of issues concerning the protection of the celestial environment. Given the nature of proposed activities under the Artemis Accords, which involves resource mining and the creation of safety zones, failure to deal with the protection of the celestial environment is not acceptable under any international legal framework.

1.6 Concluding Remarks

The Artemis Accords is the mirror image of the effort made by the United States through its domestic laws to legalise space resource mining. It is centered on the theme of commercial extraction of space resources and conferment of property rights over mined resources. One may try to draw a parallel with the Truman Proclamation of 1945, which was the first instance of asserting claims over the continental shelf. Despite some initial resistance, claims over the continental shelf gained legal recognition in due course. However, we need to keep in mind that the Truman Proclamation was made in an era where no internationally accepted norm was in existence to determine the legal status of the continental shelf. The prevailing situation in the case of outer space and celestial bodies is different as we have binding international norms applicable to them under the Outer Space Treaty. Some of the essential principles of the Outer Space Treaty have also attained the status of customary international law and *jus cogens*.

As discussed above, the provisions on space resource mining and the creation of safety zones are in direct conflict with the established principles of the Outer Space Treaty. However, these points should not be misconstrued to say that space resource mining must be prohibited or that safety zones are not required for space resource activities. With the depletion of resources on earth, space resource mining may become essential to support the increasing population. Safety zones are also required in practical terms to prevent any hazards. However, both need to be worked out at the international level with the participation of all stakeholders. An international governing body to license resource mining and designate safety zones needs to be established. The development of objective parameters under an international regime is essential to cater to the needs and interests of all. Hence, the path for multilateral negotiations needs to be reinvented, and individualistic approaches like the Artemis Accords need to be discouraged.

References

Act on the Promotion of Business Activities for the Exploration and Development of Space Resources 2021, Act No. 83 of 2021, 23 December 2021 (Japan).

Andem, M. N. (2004). *The 1967 Outer Space Treaty (1967 OST) as the Magna Carta of contemporary space law: A brief reflection*. IAC-04-IISL.3.10, 292–307. Retrieved January 31, 2025, from https://www.elevenjournals.com/tijdschrift/iisl/2004/3%20A%20General%20Convention%20on%20Space%20Law%3F/IISL_2004_047_003_006.pdf

Arato, J. (2014, March 31). *Subsequent practice in the Whaling Case, and what the ICJ implies about treaty interpretation in international organizations*. EJIL: Talk! Retrieved January 27, 2025, from https://www.ejiltalk.org/subsequent-practice-in-the-whaling-case-and-what-the-icj-implies-about-treaty-interpretation-in-international-organizations/comment-page-1/#comment-182531

Bhat, S. B. (2007). The concept of common heritage of mankind in the governance of the Moon: An insight into Article 11 of the Moon Agreement. *Legal Opus, 1*, 92–107.

Bhat, S. B. (2014). Application of environmental law principles for the protection of the outer space environment: A feasibility study. *Annals of Air and Space Law, 39*, 323–354.

Bianchet, L. (2017). *Treaty modification by subsequent practice: Analysis of the present legal framework, the expansive potential and the consequences of treaty modification by subsequent practice* (Master's thesis, Örebro University). Örebro University. Retrieved from https://www.diva-portal.org/smash/get/diva2:1189907/FULLTEXT02

Boyle, A. (2018). Soft Law in International Law-Making International Law. In *International Law* (5th ed., p. 976). Oxford University Press. https://doi.org/10.1093/he9780198791836.003.0005

Brooks, E. (1969). Control and use of planetary resources. In *Proceedings of the Eleventh Colloquium on the Law of Outer Space* (pp. 339–350).

Cheng, B. (1997). *Studies in international space law*. Clarendon.

Cheng, B. (2000). *Properly speaking, only celestial bodies have been reserved for use exclusively for peaceful (non-military) purposes, but not outer void space*. Naval War College.

Christol, C. Q. (1984). Article 2 of the principles treaty revisited. *Annals of Air and Space Law, 9*, 217.

Christol, C. Q. (1998). Important concepts for the international law of outer space. In *Proceedings of the Fortieth Colloquium on the Law of Outer Space* (pp. 73–83).

Cowing, K. (2020, July 14). Rogozin: Russia is not interested in working with NASA on Artemis (or maybe they are). *NASA Watch*. Retrieved January 30, 2025, from https://nasawatch.com/russia/rogozin-russia-is-not-interested-in-working-with-nasa-on-artemis-or-maybe-they-are/

Deplano, R. (2021). The artemis accords: Evolution or revolution in international space law? *International and Comparative Law Quarterly, 70*(3), 799–819. https://doi.org/10.1017/S0020589321000142

Fidler, D. (2020, June 2). The Artemis Accords and the next generation of outer space governance. *Council on Foreign Relations*. Retrieved January 30, 2025, from https://www.cfr.org/blog/artemis-accords-and-next-generation-outer-space-governance

Goswami, N. (2021, March 19). The strategic implications of the China-Russia lunar base cooperation agreement. *The Diplomat*. Retrieved January 30, 2025, from https://thediplomat.com/2021/03/the-strategic-implications-of-the-china-russia-lunar-base-cooperation-agreement/

Guzman, A. T. (2008). *How international law works: A rational choice theory*. Oxford University Press.

ILC. (2008). *Yearbook of the International Law Commission* 2008, Vol. II, Part 2. United Nations.

International Court of Justice. (1969). *North Sea Continental Shelf Cases (Federal Republic of Germany/Denmark; Federal Republic of Germany/Netherlands), Judgment of 20 February 1969*. I.C.J. Reports, 1969, p. 44, para. 77.

International Court of Justice. (2014). *Whaling in the Antarctic (Australia v. Japan: New Zealand intervening), Judgment of 31 March 2014*. I.C.J. Reports, 2014, 226.

Ji, E., Cerny, M. B., & Piliero, R. J. (2020, September 17). What does China think about NASA's Artemis Accords? *The Diplomat*. Retrieved January 30, 2025, from https://thediplomat.com/2020/09/what-does-china-think-about-nasas-artemis-accords/

Jia, B. B. (2010). The relations between treaties and custom. *Chinese Journal of International Law, 9*(1), 81–109. https://doi.org/10.1093/chinesejil/jmq001

Kopal, V. (1998). Outer space as a global common. *Proceedings of the Fortieth Colloquium on the Law of Outer Space*, 108–118.

Law of July 20, 2017, on the Exploration and Use of Space Resources, 2017 (Luxembourg).

Law on the Regulation of the Space Sector 2019, Federal Law No. (12) of 2019, 19 December 2019 (UAE).

Mardani, N., & Hooshmand, M. M. (2016). JCPOA; A dialectical paradigm of treaty and other international instruments. *Journal of Politics and Law, 9*(3), 70. https://doi.org/10.5539/jpl.v9n3p70

NASA. (2020). *The Artemis Accords principles for cooperation in the civil exploration and use of the moon, mars, comets and asteroids for peaceful purposes*. Retrieved from https://www.nasa.gov/specials/artemis-accords/img/Artemis-Accords-signed-13Oct2020.pdf

Neef, R. (2021). Artemis accords: A new path forward for space lawmaking? *Adelaide Law Review, 42*(2), 569–580.

Nelson, J. W. (2020). The Artemis Accords and the future of international space law. *ASIL Insights*, 24(31). Retrieved from https://www.asil.org/sites/default/files/ASIL_Insights_2020_V24_I31.pdf

Qizhi, H. (1997). The Outer Space Treaty in perspective. *Journal of Space Law, 25*(2), 93–100.

Rao, R. J. (1997). Outer space: A part of the common heritage of mankind. In V. S. Mani, S. Bhat, & V. Balakista Reddy (Eds.), *Recent trends in international space law and policy* (pp. 191–208). Lancers Books.

'Report of the International Law Commission – Seventieth Session', 30 April – 1 June & 2 July – 10 August 2018, UN Doc A/73/10.

Schingler, J. K. (2020, June 8). Imagining safety zones: Implications and open questions. *The Space Review*. Retrieved from https://www.thespacereview.com/article/3962/1

Smith, W. A. (2021). Using the Artemis Accords to build customary international law: A vision for a U.S.-centric good governance regime in outer space. *Journal of Air Law and Commerce, 86*(4), 661–700.

Steffen, O. (2022). Explore to exploit: A data-centred approach to space mining regulation. *Space Policy, 59*, 101459. https://doi.org/10.1016/j.spacepol.2021.101459

The Hague International Space Resources Governance Working Group. (n.d.). *International Institute of Air and Space Law, Leiden University*. Retrieved February 1, 2025, from https://www.universiteitleiden.nl/en/law/institute-of-public-law/institute-of-air-space-law/the-hague-space-resources-governance-working-group

United Nations. (1968). *United Nations Conference on the Law of Treaties, 26 March – 24 May 1968, 38th meeting of the Committee of the Whole*. UN Doc. A/CONF.39/C.1/SR.38.

United Nations. (1969). *Vienna Convention on the Law of Treaties*. Treaty Series, 1155, 331 (entered into force on 27 January 1980).

Chapter 2
The Inevitability of Developments and Changes to Outer Space Property Law Brought About by the Artemis Accords

Wian Erlank

Abstract After years of relative certainty about the broad outlines of property rights in outer space, which are essentially based on the non-appropriation principle on the one hand, and the obligation of perpetual supervision, jurisdiction, control, and at its essence, responsibility—regarding "space objects" on the other hand, the Artemis Accords has now started to shake things up to such an extent that it seems inevitable that outer space law must change and develop. The focus of this chapter will be on Sections 10 and 11 of the Accords, dealing with space resources, and safety zones—and the (possible) conflict that it may or may not bring with Articles I and II of the Outer Space Treaty, set against the background of (private) property law as opposed to public international law on which the current regime is based. The argument made in this chapter is that while it is possible to interpret the non-appropriation clause of Art II OST in such a way as to be an impediment to the use of Section 10 of the Accords, it is inevitable that the technological developments in outer space activities and operations will result in extraction and utilization of space resources as envisaged by the Accords. Some suggestions on how to prevent over-reach in terms of property law will follow.

Keywords Property law · Ownership right · Outer Space Treaty · Artemis Accords · Safety zones · Exclusionary rights

W. Erlank (✉)
North-West University, Potchefstroom, South Africa
e-mail: wian.erlank@nwu.ac.za

S. Bhat B., A. Variath (eds.), *Artemis Accords and Resource Mining in Outer Space*, Issues in Space, https://doi.org/10.1007/978-981-95-0152-6_2

2.1 Introduction

The creation of the Artemis Accords (hereafter the Accords) is one of the most (if not the most) exciting developments in outer space law in recent years. Its influence will, in all probability, have a huge effect on law (both public, international, and state (municipal)), policy, and definitely the effective practice of space operations and activities. In this chapter, the author will focus on some property law implications that will probably flow out of the Accords. The first thing to take note of is the fact that the Accords (at least for the moment) are, as per the U.S. Department of State, "a non-binding set of principles designed to guide civil space exploration and use in the twenty-first century" (US Department of State, 2024). As such, it is not international law nor a binding international treaty (Lyall & Larsen, 2025, 193) but could perhaps rather be seen as an agreement to interpret international law in terms of a certain set of principles (Hobe, 2021, 4). It is not international law in the sense of being universally applicable and binding on all states, such as in the case of international humanitarian law (or rules of war), which apply irrespective of whether a state has agreed to them or not. However, judging from the swift uptake of subscriptions to the Accords, it is certainly possible that it could also one day achieve the status of being considered international customary law, much like the OST (or parts thereof) is considered to be international customary law by some commentators (Ferreira-Snyman & Ferreira, 2021, 15).

Unfortunately, holding States to abide by these rules continues to prove problematic, as can be seen in the current conflicts between Russia and Ukraine, Israel and Palestine and various other countries. In this respect, the byline of the Accords provides a good précis of its aims and purpose, stating that they are "Principles for Cooperation in the civil Exploration and Use of the Moon, Mars, Comets, and Asteroids for Peaceful Purposes." These Principles have been set out in the Accords and deal thematically with a number of issues. These being addressed in Sections under the headings of: Peaceful Purposes, Transparency, Interoperability, Emergency Assistance, Registration of Space Objects, Release of Scientific Data, Protecting Heritage, Space Resources, Deconfliction of Space Activities, and Orbital Debris. The second thing to note is that while the Accords were initiated by NASA and, therefore, the USA, the end result is by no means a USA-dictated instrument. In this regard, for a particularly interesting (and accessible) discussion of the history, and creation purposes of the Artemis Accords, see (listen to) the Planetary Society's podcast called *Planetary Radio* (Planetary Society, 2022). In fact, the Accords, in encouraging international cooperation, are the result of intensive collaborative efforts between the initial parties to the agreement and, as such, should be seen as a truly international collaborative product (Planetary Society, 2022; Masson-Zwaan & Sundahl, 2021, 49. However, see Hobe, 2021, 5 for an opposing view). At the time of writing this chapter (Nov 2024) the Artemis Accords have been signed by 53 states (NASA, 2024). As such, when we discuss the question of whether States who subscribe to the Accords will be willing to implement its principles in practice, the answer is more certain (in the affirmative) due to the collaborative nature of the Accords and the fact that is not a top-down approach, imposing viewpoints and practices on members (Planetary Society, 2022). However, at the same time, a second grouping has emerged of States who plan to have a separate project to exploit the Moon, called the International Lunar Research Station (ILRS). This is led by China (via the China National Space Administration) and Russia (via ROSCOSMOS), who do not plan to join the Artemis Accords. Currently, the following 13 States have joined this project: Azerbaijan, Belarus, China, Egypt, Kazakhstan, Nicaragua, Pakistan, Russia, Senegal, Serbia, South Africa, Thailand, and Venezuela. It is speculated that we might possibly see countries join and participate in both the Accords and the ILRS, which could eventually perhaps lead to at least some form of cooperation between the two projects. This would ideally be necessary to prevent unnecessary conflict and interference with each other's projects and activities (Lyall & Larsen, 2025, 194–195), even though the OST (and other applicable space law treaties) will still be applicable as the overarching legal framework.

While the focus of a number of other contributions in this volume will be on the mineral extraction and exploitation of space resources, this chapter focuses more on some of the implications that will result from this. Extraction and exploitation will happen, and the question now is whether or not it will happen in terms that are acceptable (or at least) adoptable by the international community. Whether the practice that develops from this extraction and exploitation is reconcilable with current international space law remains to be seen. The author will also briefly deal with property law-related aspects raised in various Sections of the Accords in addition to a focussed discussion on the Sections dealing with space resources and safety zones.

2.2 The Current Property Law System

2.2.1 Background

Before we can look at the potential disruption the Accords may bring about to outer space law, we should start with a quick overview of the current system. As alluded to above in the introduction, the current interpretation of property rights (and, as such, the legal framework on which this is based) is essentially one based on public international law. As such, it focuses on and is informed by the negotiations, treaties, agreements, rules, interactions, and actions of States—and much of these were heavily influenced by the political climate created by the cold-war between the United States of America (USA) and the Union of Soviet Socialist Republics (USSR) (Lyall & Larsen, 2018; Blount, 2011; Erlank, 2015b). During this period we also saw the aspirations of both countries turn to outer space and as such the rivalry resulted in the space-race, culminating in the USA "winning" by landing a man on the Moon as part of the Apollo 11 mission in 1969. However, we need to take a step back and realise that the race to the Moon could also have been a possible race to acquire new territory for whichever State reached it first. As such we see that the creation and adoption of the Outer Space Treaty preceded the eventual Moon landing, and as such provided the international legal framework for what rights would vest in the party who won the race. While there are numerous other legal aspects to reaching (and landing on) the Moon, for purposes of property law, the question of the possible acquisition of territory (in the form of acquiring territorial sovereignty, and as such, claiming the Moon as their (sovereign) property), remains central to property law in outer space. To put this in a slightly different perspective, due to (amongst other aspects) the uncertainty of who was going to win the space race and be the first to land a (hu)man on the Moon, parties were willing to come to an agreement that the generally applicable principle of *terra nullius* will not apply to outer space (Lyall & Larsen, 2018). As such, outer space, the Moon and other celestial bodies were deemed not to be *res nullius* (things which belong to no-one and are capable of being appropriated in terms of ownership) (Erlank, 2023b, 203–204), but rather *res communis omnium* (things which belong to everyone and as such are not capable of being appropriated in terms of (private) ownership) (Freeland, 2013; Erlank, 2016a; Tjandra, 2021, 374). The other Roman law-based term that is used interchangeably with *res communis omnium* is *res omnium communes*—in most of the space law literature this concept is just referred to as *res communis*. The fear, of course, was that if such an agreement was not reached, whoever got to the Moon first, could, by means of *occupatio* (as a form of original acquisition of ownership) (Erlank, 2016a) claim the Moon for themselves and as such, in terms of territorial sovereignty (for public international law), or ownership rights (for private law), exclude all others from the Moon, or portions thereof (Erlank 2023b, 203; 2025). (Original acquisition of ownership denotes a person receiving or acquiring the right of ownership without the existence or co-operation of a previous holder in title. Practically speaking, in most cases, this refers to becoming the owner of a thing that was not previously owned by anyone else, or, to which the previous owner has lost their right of ownership due to a.o. losing control over the object (such as with wild animals), or by abandoning their ownership right). This would have been untenable and frankly, counterproductive for mankind's endeavours into outer space.

It is interesting to note that the symbolic form of claiming territory for one's own (by States), whether it be *terra nullius* (land belonging to no-one) or freshly conquered from some other State, often goes hand in hand with the planting of a flag (Erlank, 2016b, 6). As such, the USA was extremely careful to ensure that their planting of the US flag at the landing site of the Eagle lunar lander, was not construed as an act of appropriation, contra to the agreed upon terms of the OST. In fact, the USA created discrete legislation prior to the Moon landing to make it explicitly clear that the planting of the US flag was in fact, symbolic, but only in terms of mankind's achievement of reaching the Moon, and definitely not in terms of appropriating ownership of the Moon (Lyall & Larsen, 2025).

2.2.2 *Ownership/Or the Lack Thereof in Terms of the OST*

Having thus discussed a very quick overview of the circumstances and climate for international treaty-making, we can now turn our focus to what is, in fact, contained in the OST regarding the acquisition of property rights. While there are a number of sections in the OST that have a bearing on property and property relations, the most predominant one is Article II, which contains the so-called non-appropriation principle (Blount, 2011; Freeland, 2013). The text of this principle is relatively straightforward, and at least on initial reading it is quite clear that the acquisition of outer space, the Moon and other celestial bodies is absolutely forbidden. Just to make certain that this prohibition is not misinterpreted, a number of examples of methods of appropriation are specifically mentioned—and then a catch-all phrase is included. Article II of the OST states: "Outer space, including the Moon and other celestial bodies, is not subject to national appropriation by claim of sovereignty, by means of use or occupation, or by any other means."

In other words, national appropriation is prohibited, and this also applies to circumstances where this could be done by claiming sovereignty (over territory), or using such territory, or by means of occupation (original acquisition of property) and or, just to make sure that all other scenarios are covered, ... by any other means. As such, at the moment, it is unequivocally clear that in terms of outer space law, the acquisition of celestial real estate (or any of its iterations such as heavenly bodies, comets, asteroids, planets etc.) by States is prohibited (Erlank, 2025). This applies to appropriation for purposes of vesting territorial sovereignty as well as for purposes of obtaining private ownership (for those States where it is possible for the State to have private ownership). What is not clear, at all, is whether private citizens, companies, or non-governmental organisations are also prohibited from appropriating such heavenly bodies (or portions thereof) (Pop, 2010, 65; Erlank, 2015a, 1767). Upon analysis it seems that the prohibition against acquiring property rights (and ownership specifically) is explicitly aimed at States, and that, by clear omission, this does not seem to be applicable to private actors as set out above. However, consensus has not been reached about this point, and opposing opinions are strongly motivated on both sides of this debate (see, for example, Erlank, 2015b; Freeland, 2013; Pop, 2010, 65–68).

It is important to keep in mind that this absolute prohibition placed on States with regard to acquiring property rights to celestial real estate, does not apply to all property in outer space (Erlank, 2025). In fact, there is a clear obligation placed on States that not only affirms ownership rights to property in Outer Space, but also highlights the responsibilities that come with ownership of such property (Lyall & Larsen, 2018). While ownership is regarded as the most powerful and complete right that one can obtain with regard to property, often being called an absolute right, it also comes with responsibilities or obligations that an owner has to comply with, or restricts/burdens the way in which an owner can use their property (Erlank, 2025). One can even go so far as to agree with Pop (2010, 71) that the concept of private property is "an eroded concept ... [where] the 'owner' has [e]volved into a mere 'privileged user' of a certain object". Using ownership of a car as an example, this includes the obligation to pay taxes, have a driver's licence, to follow the rules of the road such as speed limits, ensure the vehicle is roadworthy, avoiding using the vehicle to cause harm to others, etc. Hohfeld described this interplay between rights and obligations as rights on the one hand balanced against correlative duties on the other hand (Hohfeld, 1917).

As mentioned above, original acquisition of ownership denotes a person receiving or acquiring the right of ownership without the co-operation or existence of a previous holder in title. Practically speaking, in most cases, this refers to becoming the owner of a thing that was not previously owned by anyone else, or, to which the previous owner has lost their right of ownership due to a.o. losing control over the object (such as with wild animals), or by abandoning their ownership right. In this case, I am referring to the specific category of things called "space objects". In terms of outer space law, the term "space objects" should not be confused with the generic descriptive term of an object in space, but rather should be understood to be a discretely (but unfortunately not clearly) defined category of objects (Lyall & Larsen, 2018; Erlank, 2025). Unfortunately, the definition of a "space object", while providing some guidance, is mostly in the form of a circular argument that comes down to "you'll know it when you see it". The definition of a "space object" is derived from the similar provisions contained in the Registration Convention (Art 1(b)) and the Liability Convention (Art 1(d)), which states that a space object has the following meaning: "The term "space object" includes component parts of a space object as well as its launch vehicle and parts thereof."

Essentially, this means that anything launched into outer space by a launching state is considered to be a space object. In terms of the Liability Convention, objects on the surface of the Earth, or in the atmosphere can also be categorized as space objects if they have been launched to or into outer space, or have been destined to be launched into outer space. However, the focus of this article is just to compare naturally occurring objects in outer space (which fall foul of the non-appropriation principle) against those mostly human-made objects launched into outer space or crafted in outer space, to which ownership accrues (Erlank, 2025). While the nuances of what is and is not considered to be a space object, is the subject of many other studies, for purposes of this chapter, it will suffice to place this against the naturally occurring heavenly bodies and other objects found in outer space (Erlank, 2025).

While it is clear that heavenly or celestial bodies and outer space in general is not appropriable by States, space objects are definitely owned by or deemed to be owned by those (launching) States, responsible for placing such space objects in outer space. While ownership of space objects initially, by default vested in States, since they were the only participants in outer space activities, a large number of space objects are now launched, owned and operated by private companies. However, since launching states remain liable for the actions of such private companies falling within their jurisdiction, as well as for the authorization, supervision and control over such space objects, these states are still regarded as the owners of such space objects in terms of space law (Pop, 2010, 60). As such Lyall and Larsen (2018) argue that one of the reasons why a space object needs to be registered in terms of the Registration Convention is because "... entry on that register is important for grounding responsibility, for 'ownership', for the exercise of control and in the worst case, for liability." This mandatory ownership is derived from the obligation placed on Launching States by the OST (Art VI and VIII) to authorize, supervise and exert control over all the space objects for which they are responsible (Lyall & Larsen, 2018). More specifically in terms of Art VI, a State is made responsible for the activities of itself and its citizens in outer space, it needs to authorise and continually supervise all activities of non-governmental entities; and such a State retains jurisdiction and control over all space objects (and personnel) (Erlank, 2023a, 167). With these responsibilities also comes the liability for damage caused by such space objects. Interestingly enough, unlike in the case of many types of things owned on Earth, it would seem that it is currently impossible for a Launching State to unilaterally abandon ownership of a space object—since liability for such object cannot be erased by such an act (Lyall & Larsen, 2025). The launching State remains liable for damages caused by such space object ad infinitum. While launching states remain liable for their space objects and cannot unilaterally fully abandon their ownership of these objects, there are mechanisms to transfer ownership from one party to another (since this discussion falls outside the scope of this chapter, see in general Lyall & Larsen, 2018).

To summarise, the position in this section focused on the prohibition contained in Article II of the OST that applies to States, and does not necessarily preclude private appropriation of celestial real estate. It was also highlighted that while this prohibition against appropriation of celestial bodies exists, states do have ownership of all of their space objects.

2.3 The Artemis Accords

2.3.1 Introduction

As mentioned at the beginning of this chapter, the Accords deal thematically with the following Sections: Peaceful Purposes, Transparency, Interoperability, Emergency Assistance, Registration of Space Objects, Release of Scientific Data, Protecting Heritage, Space Resources, Deconfliction of Space Activiteis, and Orbital Debris. Each of these will be discussed in turn, (some in more depth than others), as they touch on property law issues and then tying into the previous discussion of the history (and currently applicable) outer space legal framework for property law. Interestingly enough, in the preamble to the Accords, only four of the five main space treaties are acknowledged, and (unsurprisingly) the Moon Agreement is specifically left out. While this is unsurprising because the Moon Agreement is generally viewed as a failed Treaty, it is interesting since some of the ideas from the Moon

Agreement clearly informed the Accords, especially the section on Space Resources. In this section I will briefly note some property implications stemming from some of the Sections, and will then delve into greater depth concerning the Sections dealing with Space Resources and Deconfliction (Safety zones).

2.3.2 General Commentary on Sections of the Accords

Section 3: Peaceful Purposes, highlights the fact that the Accord's cooperative activities should be used exclusively for peaceful purposes and in accordance with international law. (NASA, 2024) highlights the fact that such cooperation should "enhance peaceful relationships between nations … consistent with the Outer Space Treaty." Section 4 on 'Transparency', highlights the need for the exchange and dissemination of information about policies, plans as well as the sharing of scientific information on a good faith basis. Section 5, 'Interoperability' focuses on the practical need for use, development and maintenance of standards for interoperability. The US Department of State (2024) notes that: "[w]orking in the space environment is incredibly challenging. Interoperability enhances the potential for space exploration that is safe and robust among cooperating nations." Especially due to the challenging environment, which is inimical to human life, interoperability of standards and equipment is essential, with standards for communication and interoperability of such things as airlocks and other common and essential exploration and life-sustaining infrastructure structures and systems being essential for human activities in outer space. This means that there will possibly be more collaboration on technological development rather than a focus on proprietary technology. Collaboration on the development of such new technologies should speed up the development of new and better technologies, with a clear benefit for all parties, rather than working separately in silos of knowledge, which could hamper, rather than support, exploration and utilisation of outer space.

Section 6, 'Emergency Assistance', acknowledges the Rescue Agreement and makes an undertaking of Signatories making reasonable efforts to render necessary assistance to personnel in outer space who are in distress. While the commitment here is luke-warm at best due to the uncertain nature of what a "reasonable effort" is, the fact that the Accords to not speak about "Astronauts" but rather "personnel" is indicative of the fact that the technicalities of who is and is not eligible for assistance under the Rescue Agreement is more clear (Martin & Freeland, 2022, 275–277). However, the use of "persons" rather than "personnel" could have made this even clearer. This also ties into the interoperability standards above, making assistance and rendering assistance easier if infrastructures are interoperable.

Section 7 'Registration of Space Objects' deals with the important requirement of registration of space objects, which is important from a property perspective for the allocation of ownership rights, responsibilities and liability. Since Signatories are going to cooperate in space activities, it is logical that certainty about the legal status of any particular space object is achieved and traceable. Since registration of space objects (even by signatories to the OST and the Registration Convention) is inconsistent, the Accords will possibly ensure better adherence to the important principle, including the need for non-parties to register their space objects. However, as noted by Hobe (2021, 6), this requirement might fall foul of the *pacta tertiis* rule unless one argues that the duty to register has acquired the status of customary international law. Since more and more space objects will invariably be created in outer space, including from material mined, extracted and beneficiated in outer space, this requirement is paramount for keeping track of the ever-increasing volume of new space objects—not launched from or originating from Earth (Erlank, 2025). This is also evident from the focus of the US Department of State on the mitigation of risk of harmful interference, who note that "... [a]ppropriate registration of space objects can help to mitigate risk of harmful interference." (US Department of State, 2024).

Section 8 on 'Release of Scientific Data' deals with the commitment of open sharing of scientific data, which ties into the goals of sharing the benefit of space exploration and use, as well as the requirements of interoperability. While noble in its intention, the exclusion of the need to share proprietary and or export-controlled information, as well as not being applicable to private sector operations (unless conducted on behalf of a Signatory), makes this Section of little practical value. It does bode well for ownership, intellectual property rights and commercialisation of such data.

Section 9 'Preserving Outer Space Heritage' acknowledges the importance and need for the preservation of historically significant sites and artifacts. "The Signatories intend to preserve outer space heritage, which they consider to comprise historically significant human or robotic landing sites, artifacts, spacecraft, and other evidence of activity on celestial bodies in accordance with mutually developed standards and practices." (Section 9.1, Accords).

The identification of such sites and artefacts will, of course, need some deliberation and negotiation to come to a consensus on what should be included and what should be excluded. In all probability, the recognition of landing sites and areas of activity could lead to the creation of areas that are either no-go zones or else removed from general access and the ability to use and exploit such areas. On the one hand, this impedes the principle of free access, but on the other hand, it can be a noble attempt to preserve certain pristine or significant areas from interference. Since one man's junk is another man's treasure, negotiations about which areas and space objects will be included in such a list could be problematic, especially if one considers the current practice that the space object of another State is sovereign property and untouchable, even if such object is non-functional, useless or hazardous. As such, it is possible that certain States may claim useless and or irrelevant space objects as "artefacts" or the places where they are located as historical sites—with the result that they effectively block others from accessing and utilising those areas, possibly for strategic purposes.

Sections 10 'Space Resources' and 11 'Deconfliction of Space Activities' will be discussed in more detail below. Section 12 'Orbital Debris' is exceptionally important from a property perspective since it deals with the need to mitigate the proliferation of orbital debris by limiting the generation of "new, long-lived harmful debris", as well as for the post-mission disposal of spacecraft and other space structures. This is done by highlighting the need to plan for and implement these strategies as a part of the mission planning process; and where there are cooperative missions, determining who will take the primary responsibility for this. The US Department of State (2024) notes that "[p]lanning to mitigate for orbital debris, as well as disposing safely of spacecrafts [sic], is critical to maintaining a safe environment in space and operating in space sustainably." While this is not one of the longest sections of the Accords, it could very well be one of the most important. The proliferation of orbital debris around the Earth, and the threat of a looming cascade-effect of proliferation of space debris, cutting of acces to outer space, remains a real and increasing problem. As such, it logically follows that stepping up activities in outer space as envisaged in the Accords will result in even more orbital debris around the Earth, as well as around the Moon and between the Earth and the Moon, and eventually further. This is, of course not a sustainable practice, even though recent efforts to address this has started to show some promising buy-in by certain space actors.

> "1. The Signatories commit to plan for the mitigation of orbital debris, including the safe, timely, and efficient passivation and disposal of spacecraft at the end of their missions, when appropriate, as part of their mission planning process. In the case of cooperative missions, such plans should explicitly include which Signatory has the primary responsibility for the end-of-mission planning and implementation.
>
> 2. The Signatories commit to limit, to the extent practicable, the generation of new, long-lived harmful debris released through normal operations, break-up in operational or post-mission phases, and accidents and conjunctions, by taking appropriate measures such as the selection of safe flight profiles and operational configurations as well as post-mission disposal of space structures." (Section 12, Accords).

It is heartening to see that the Accords make provision for the planning and execution of preventative and clean-up activities to ensure sustainable operations. This also ties into the need to prevent and mitigate harmful interference with one another's activities. Cleaning up, dismantling and disposing of spacecraft, debris, and structures will also ensure that others are not excluded from access to unused or previously occupied areas in outer space and on celestial bodies in terms of my argument about effective control, which will be discussed in more detail below.

2.3.3 Space Resources

Section 10 on 'Space Resources', unsurprisingly, deals with the question of access to extraction and utilisation of space resources. Since this is the section of the Accords with the most importance for property law, it is important to take a look at the complete text first before analysing it and its implications for property rights in outer space in more detail.

> "1. The Signatories note that the utilization of space resources can benefit humankind by providing critical support for safe and sustainable operations.

> 2. The Signatories emphasize that the extraction and utilization of space resources, including any recovery from the surface or subsurface of the Moon, Mars, comets, or asteroids, should be executed in a manner that complies with the Outer Space Treaty and in support of safe and sustainable space activities. The Signatories affirm that the extraction of space resources does not inherently constitute national appropriation under Article II of the Outer Space Treaty, and that contracts and other legal instruments relating to space resources should be consistent with that Treaty.
> 3. The Signatories commit to informing the Secretary-General of the United Nations as well as the public and the international scientific community of their space resources extraction activities in accordance with the Outer Space Treaty.
> 4. The Signatories intend to use their experience under the Accords to contribute to multilateral efforts to further develop international practices and rules applicable to the extraction and utilization of space resources, including through ongoing efforts at the COPUOS." (Section 10—Space Resources, Accords).

As background to the discussion to follow, it is illuminating to note NASA's explanatory note on this section, which highlights the stated goal of compliance with the OST, while also providing the justification for extraction of use and resources as being "critical to support safe and sustainable space exploration and development."

> "The ability to extract and utilize resources on the Moon, Mars, and asteroids is critical to support safe and sustainable space exploration and development. The Artemis Accords reinforce that space resource extraction and utilization can and should be executed in a manner that complies with the Outer Space Treaty and in support of safe and sustainable space activities." (NASA, 2024).

This is echoed by the US Department of State's summary, which states that: "The utilization of space resources should be done in a manner that complies with the Outer Space Treaty, can benefit humankind and is critical to sustainable operations." (US Department of State, 2024). Two things are crystal clear in this portion of the Accords. The first is that there is an unambiguously stated goal of extracting and utilizing space resources. The second is an explanation/motivation for the legality of this in terms of outer space law as being "critical" for safe and sustainable exploration and activities. This is a logical necessity if one considers that it is not only impractical but severely limiting to undertake space exploration and development by only using resources and supplies that originate on Earth. Do note that there is no reference to the commercialisation of such extraction practices, nor of the commercialisation and sale of such extracted resources. While it is easy to accept that extraction and usage for critical and necessary resources connected with general use, exploration and development in outer space (*in situ*) is easy to justify, cynical commercial exploitation for the profit and benefit of a restricted privileged group, to the exclusion of the rest of society, is not such an easy argument to make, and does raise questions about the fact whether such actions are really for the benefit of humankind. It also does not address the pervasive question of how benefits from such extraction and usage will be shared. The need for a regime to share benefits from space mining and exploitation remains relevant (Xu & Su, 2022, 3). Irrespective of one's views about this; it is good that the Accords do provide a forum and template for coordinating the inevitable extraction and exploitation of celestial bodies, even if it is initially only focused on the specific component of space resources. The fact of the matter is that even if most States agree to the strict interpretation of the non-appropriation principle as to exclude the extraction and use space resources, economic interest will override this, and could lead to a "finders-keeper", "winner-takes-all" situation where those who have the finance, means, technological prowess and inclination to get to space resources and start extracting and exploiting the same, will be able to protect their financial and operational interests—perhaps by military (private or State) means (Erlank, 2017, 83).

Furthermore, as to the legality of extracting and utilizing space resources, the Signatories have come out with a clear message on how they are going to interpret the ambiguity regarding the right to use as set out in Article I, and the non-appropriation principle as set out in Art II of the OST. This is done by means of the following phrasing: "The Signatories affirm that the extraction of space resources does not inherently constitute national appropriation under Article II of the Outer Space Treaty" (S. 10, Accords). While the message is clear, that the Signatories are interpreting Art II of the OST as to allow for the extraction of space resources, it could, perhaps, have been written a bit stronger by leaving out the word "inherently". As it stands, it leaves the door open for instances where the extraction of space resources could, in fact, constitute national appropriation, particularly in the scenario where one argues that a State can appropriate resources (or territory) via the proxy of their private citizens, private companies or non-governmental organisations (Erlank, 2025).

From a property perspective, it is interesting that the Accords clearly distinguish between extraction and utilization. In terms of Art 1 of the OST, the freedom of "exploration and use" is mentioned twice, first in the context of the "benefit and in the interests of all countries", and second in the context of being "free for exploration and use by all States". In this case, use is therefore interpreted in the Accords to mean that the freedom of use accorded by the OST, includes the right to exploit, make use of, and extract. Of course, using such space resources, even without extracting, by, for example, constructing a lunar base on and or below the surface of the Moon, will also be in accordance with the permission to use as set out in the OST.

The act of extracting space resources in the Accords is directed specifically at the surface and subsurface of the Moon, Mars, comets and asteroids, but not necessarily excluding other celestial bodies. The Accords also make use of the phrase "space resources" rather than celestial bodies to denote that such "resources" are differentiated from the heavenly body as a whole. While the current legal status of space resources is uncertain (Xu & Su, 2022, 2), Svec (2022, 2), amongst others, argues that the non-appropriation principle neither excludes the exploitation of space resources nor are they explicitly included under the common heritage of mankind. As such, since the extraction and utilisation of "resources" are not prohibited in the OST, the fact that such extraction is done on or from a heavenly body should not prove to be a problem. How then is it possible to conceptually sever the extraction of resources from the prohibition to appropriate the whole of a heavenly body? As mentioned above in the background to the current property regime in outer space, the OST has designated the whole of outer space, including the Moon and other celestial bodies, as the common heritage of mankind and, therefore, being *res communis* (Pop, 2010, 60). This means that, for purposes of property law, outer space (and all the heavenly bodies in it) falls outside of commerce—*res extra commercium* (Erlank, 2016a) and cannot be bought or sold. Pop (2010, 65) argues that the distinction between full ownership and *res communis* is the latter's "lack of marketability",—which highlights the fact that *res communis* are also *res extra commercium*. If outer space was, in fact, *res nullius*, then there would not be any problem for extracting resources (or minerals) from heavenly bodies, since the fact that they don't belong to anyone would allow for the extraction, as well as the possibility to acquire ownership of the resources and possibly the property from which it was extracted.

Original acquisition of such resources will mean that the person extracting them, will become the owner of the resources by means of *occupatio*. As to the international position on this, Pop (2010, 69) argues that Russia set the stage for the acceptance of this viewpoint by commercialising lunar material that was retrieved by an earlier Soviet lunar probe. In fact, he goes further by arguing that since: "… no objections were voiced by third [party] States, it can be stated that, as an attribute of ownership, the right to commercialise extraterrestrial material has entered into customary international law" (Pop, 2010, 69; Tjandra, 2021, 386). The US further paved the way for international acceptance of the practice of commercializing (extracted) space resources by adopting the *Commercial Space Launch Competitiveness Act* in 2015 (Erlank, 2017, 73–75), which was then followed by Luxembourg's *Law on the Exploration and Use of Space Resources* in 2017, the UAE's *National Space Law*, in 2019 and followed by Japan's *Law Concerning the Promotion of Business Activities Related to the Exploration and Development of Space Resources* in 2021 (Tronchetti & Liu, 2021, 2–3; Svec, 2022, 1–2). It seems that India will also allow for space mining by appropriately licenced non-governmental entities (Lyall & Larsen, 2025, 197). Alternatively, one could invoke Lockean labour theory (sweat of the brow) as a reason for acquiring ownership of the extracted resources since one has expended one's time and effort into extracting and acquiring the resources (Pop, 2010, 69; Erlank, 2016b). However, as mentioned above, outer space and all of the celestial bodies in it are not *res nullius*, and as such, cannot be appropriated (by States). How does one then justify allowing anyone (private or State) to acquire ownership of the resources if the place where they are extracted from cannot be owned? In this case, we once again turn to Roman Law and see that things that were deemed to be *res extra commercium*, could in fact, in certain instances, become appropriable and, as such, revert to *res intra commercio* (Erlank, 2016b). This happened when one was able to (by means of one's effort) exert control over a portion of the whole. For example, one could not own the sea, but if you placed some seawater into a jug, you could own the seawater contained in that jug.

2.4 Deconfliction of Space Activities: The Safety Zone

2.4.1 Possession and Exclusion and Property in Outer Space

Tying into the discussion of the extraction of space resources is the fact that States and private actors will soon be building structures on the Moon (and eventually on other celestial bodies). Here, reference is made to some of the practicalities and consequences that will flow from these actions. Before coming to the envisaged "safety zone", we need to acknowledge that any structure placed onto the surface of the Moon, or embedded into or below the surface of the Moon (or other celestial body) will have certain factual consequences. The most important of these from a property viewpoint is that the building, landing, placement, embedding or attaching of such structures (which are also deemed to be space objects—as discussed earlier) will lead to the possession or occupying of such area. Without going into the legality of such possession, the fact of the matter is that such factual possession or occupation means that all other States or space actors will be automatically excluded from occupying or possessing that same space (Tjandra, 2021, 387). No two space bases/spacecraft/lunar rovers can occupy the same three-dimensional (space) location at the same time. As such, use via possession of such space by one State or Actor automatically excludes all others from using the same space. This results in a *de facto* situation where physical possession is 9/tenths of the law and especially in terms of exclusion, will result in acquisition of possession as a property right, and could certainly almost be equated to ownership. Another question that flows from this automated act of physical possession, is if such placement/landing/building/embedding could be viewed as an act of attachment or superficies—which results in the loss or acquisition of ownership of either the space object (base/craft) or the portion of the celestial real estate where it is located (Erlank, 2025).

In such cases, the Moon Agreement provided some certainty in Article 11 (3), which made it clear that the act of placing structures and personnel on or below the surface does not create any rights of ownership. "The placement of personnel, space vehicles, equipment, facilities, stations and installations on or below the surface of the Moon, including structures connected with its surface or subsurface, shall not create a right of ownership over the surface or the subsurface of the Moon or any areas thereof" (Art 11 (3) Moon Agreement).

Since the Moon Agreement is specifically NOT referenced or acknowledged in the Accords, one can only wonder if such acts of placing, installation or connection could in the future, create rights of ownership (by means of use or occupation) for at least private actors and perhaps even indirectly for the States themselves (via proxy) (Erlank, 2025). On that point it is important to remember that it is clearly stated in the long title of the Accords (noting civil exploration) that the Accords and principles are aimed at civil space activities and civil space agencies: "The principles set out in these Accords are intended to apply to civil space activities conducted by the civil space agencies of each Signatory." (Section 1, Artemis Accords).

The "civil" part of the space activities and agencies is not defined or described in detail, and obviously includes the well-known national space agencies (such as NASA) and supra-national space agencies, (such as ESA), although the argument that these agencies are civil (in the sense of not being a part of the State) and as such non-governmental or private will be difficult to make. The other differentiation could be between civil(ian) and military. However, when taking the context of the background and aims of the Artemis program, together with the support for citizens (civilians) to extract and use space resources, it is clear that the Accords will apply to private, non-state activities as well, and private (civil) actors such as SpaceX or

Blue Origin could possibly be included in the understanding of "civil space agencies". On the one hand it makes sense, since each State still remains responsible (and liable) for their own activities or those which they authorize and supervise as "Launching State", but on the other hand, if it is argued that all private (and non-state) actors need to be included in the Accords under the supervision of a civil space agency such as NASA, it could result in a counter-intuitive and counterproductive situation. This aspect should be more clearly defined in a future version or amendment of the Accords.

The reason why this is important, apart from the questions raised above, is due to the uncertainty about the application of the OST's non-appropriation principle to private or non-state actors. As discussed above, while the application of the non-appropriation principle to States is crystal clear at the macro-level (whole celestial bodies, and outer space in general), at the micro level (use of, extraction and exploitation of space resources—especially if such resources are not self-regenerating (*non renascentia)* and can be used up (*res consumptibles*)—the position is by no means clear. In Roman law, minerals were divided into those that were considered to be self-regenerating or renewing (renascentia) and those that did not renew themselves (*non renascentia*). However, it is important to note that the Romans viewed some minerals (incorrectly) as self-renewing, but modern applications could still include such examples as salt and sand deposited by tides or rivers. (Van der Merwe, 1979, 59). Since it is difficult to view extraction of minerals or space resources purely in terms of use (ie not-consuming) the reality of the matter is that most of the use in the near future will comprise of exploitation by consumption and using up the space resources. If one were to follow the narrow interpretation that extraction (mining) and consuming or using resources will fall afoul of the non-appropriation principle for States, commercial extraction, consumption (and selling) of such space resources should also not be allowed for civilian space agencies such as NASA. Especially if States try to artificially remove their "civilian" space agencies and activities from the oversight and application of the OST. At the same time, it is my view that the use, exploitation and consumption of space resources for purposes of non-commercial space exploration and associated development should pose no problem for States, and as such, not fall afoul of the non-appropriation principle. The distinction is, however, one that will need more study. Returning to the fact that the non-appropriation principle most probably does not apply to private, non-state actors, the extraction, use, consumption and (acquisition of ownership) for commercial purposes should logically follow.

2.4.2 Section 11: Deconfliction of Space Activities with Safety Zones

The Accords, in Section 11, dealing with Deconfliction of Space Activities, introduce the concept of so-called safety zones, which will apply to Signatories. This is done against the background of especially the principles of due regard and the prevention of harmful interference with the activities of other States as set out in the OST. The reasoning provided for the creation of safety zones, which are both logical, practical and well-motivated (at least on an initial reading of the text) is introduced as a method of or for the purpose to prevent or de-escalating potential and actual conflicts that Signatories might experience when undertaking activities on the Moon or other celestial bodies. It comprises two broad components. On the one hand, Signatories will attempt to refrain from intentional actions that may create harmful interference and provide each other with information about the location and nature of their activities if they believe that activities may result in harmful interference or pose safety risks to one another's activities. "The Signatories commit to provide each other with necessary information regarding the location and nature of space-based activities under these Accords if a Signatory has reason to believe that the other Signatories' activities may result in harmful interference with or pose a safety hazard to its space-based activities." (Section 11.6, Accords).

On the other hand, the second component deals with the establishment, location, size, scope and ending of safety zones. While the volume of the text in the Accords dealing with these aspects concerning safety zones is quite large, it, unfortunately, manages to provide very little actual guidance on how these safety zones will operate or what their actual nature will be. The Accords define a safety zone (loosely) as: "A safety zone should be the area in which nominal operations of a relevant activity or an anomalous event could reasonably cause harmful interference." (Section 11.7, Accords). The following principles are set out concerning such safety zones:

> "(a) The size and scope of the safety zone, as well as the notice and coordination, should reflect the nature of the operations being conducted and the environment that such operations are conducted in;
>
> (b) The size and scope of the safety zone should be determined in a reasonable manner leveraging commonly accepted scientific and engineering principles;
>
> (c) The nature and existence of safety zones is expected to change over time reflecting the status of the relevant operation. If the nature of an operation changes, the operating Signatory should alter the size and scope of the corresponding safety zone as appropriate. Safety zones will ultimately be temporary, ending when the relevant operation ceases; and
>
> (d) The Signatories should promptly notify each other as well as the Secretary-General of the United Nations of the establishment, alteration, or end of any safety zone, consistent with Article XI of the Outer Space Treaty." (Section 11.7, Accords).

As such we do know that the size and scope of the safety zone should not be arbitrary, but be reflective of the nature and environment where operations are conducted. It should be reasonable with reference to or measured against "accepted scientific and engineering principles". The nature and lifetime of safety zones is fluid and can change over time in reaction to the requirements and status of the operation and very important, the safety zone will be temporary and end when the connected operation ends. While this seems as if it provides some firm guidelines, in reality it raises a number of questions and leaves room for exploitation and circumvention (Hobe, 2021, 9–10).

From a property perspective, and keeping to the focus of this discussion on the non-appropriation principle and the question of property rights, the creation and maintenance of safety zones will in all probability lead to large areas of exclusive use for the Signatory maintaining such safety zone, additionally the safety zone, will most probably effectively be an exclusion zone, strictly controlled by the Signatory undertaking operations on the associated site or location. As such, the occupation/possession of the base or installation will already exclude others from the area, and the associated safety zone will expand the reach of this exclusive use or exclusion zone and as such prevent others from using the same area during the operation of the safety zone. This is possibly contra to Art I and II of the OST (Hobe, 2021, 9). This is notably different form the approach in the Moon Agreement where Article 9 limits the extent of the area used by a State, to that which is required for the needs of the station, and stations should not impede free access to the Moon for other States.

> "1. ... A State Party establishing a station shall use only that area which is required for the needs of the station ...
>
> 2. Stations shall be installed in such a manner that they do not impede the free access to all areas of the Moon of personnel, vehicles and equipment of other States Parties conducting activities on the Moon ...." (Art 9, Moon Agreement)

In the Accords, specific mention is made to the fact that the principle of free access to all areas of celestial bodies is respected: "The Signatories commit to respect the principle of free access to all areas of celestial bodies and all other provisions of the Outer Space Treaty in their use of safety zones." (Section 11.11, Accords). However, the reality is that this principle is tempered by reference to the necessities of protection of personnel, equipment and operations, as well as the inference that protection of proprietary and export-controlled information could have an effect on when, how (much), and possibly if, information on safety zones is made public.

> "The Signatory establishing, maintaining, or ending a safety zone should do so in a manner that protects public and private personnel, equipment, and operations from harmful interference. The Signatories should, as appropriate, make relevant information regarding such safety zones, including the extent and general nature of operations taking place within them, available to the public as soon as practicable and feasible, while taking into account appropriate protections for proprietary and export-controlled information." (Section 11.9, Accords)

I foresee that this focus on the protection of personnel, equipment, operations and proprietary and export-controlled information will be used to enforce exclusion from safety zones as a general rule and not as an exception. This will invariably have a restrictive effect on the principle of free access, and will possibly lead to the eventual development of a new framework for property rights (which might not necessarily include ownership) in the future.

Again, I would like to focus on the practical property issues that this will raise. As with the situation discussed above about the automatic exclusion of others by the placement of installations /bases or other personnel and space objects on the Moon or another heavenly body, the extension of a so-called safety zone(s) around such bases/installations or other sites of operation, as well as the very real possibility that such safety zones will be used to effectively or essentially exclude other parties from the area (Vertadier & Schingler, 2023, 200), will result in a norm where the actors undertaking a particular operation connected to a safety zone will definitely acquire *de facto* property rights to the area (at least in terms of possession and the right to exclusive use) or ownership in all but name. This is not a new situation, as it has been argued that the actual acquisition of ownership to or on (portions of) heavenly bodies might be achieved in all but name through other avenues such as physical possession, exploitation, licencing, concessions, etc. (Erlank, 2016b).

2.4.3 Allocating Property and Exclusionary Rights: A Possible Solution

Two things in this Section of the Accords are laudable: the fact that the existence of safety zones should be limited in extent and in time to that which is necessary. Ideally, one can hope that the same will apply to the size and length of placement of structures /installations/stations which form the centre of these safety zones. If one keeps in mind the fact that while one station is occupying a specific site on the Moon, or any other celestial body, no one else can occupy or use that same area, the indefinite possession of such area of the celestial body could be construed as having *de facto* ownership of that area in all but name (Erlank, 2017, 83). As such I have argued (Erlank, 2015b, 2512) that irrespective of the name given to the use and occupation of a portion of a celestial body, physically by the structure, or operationally in terms of an area surrounding such structure, there should be clear rules about the extent of area of exclusive use as well as the legal status of such use and occupation. This should be based on the principle or requirement of effective control (Erlank, 2017, 73). The argument is that rather than deny the existence or creation of property rights which might present themselves from especially the factual situation of possession (and exclusion) as described above, one should accept that in certain circumstances, property rights to objects in space (including resources, extracted and *in situ*, as well as portions of heavenly bodies) should be recognised (within strict limits); together with the development of an "appropriate legal framework, rather than denying the existence of such rights and thereby stifling innovation, the development of space-related technology and investment in space exploitation and travel" (Erlank, 2017, 84).

In essence, one can only acquire a right to such a portion and size of such portion of a celestial body that one can exert effective control over, and that is necessary for the effective use or exploitation of such portion. This will prevent territorial acquisition for the reason of national appropriation as foreseen in Art II of the OST and prevent a situation of under-exploitation and underutilisation of the (commons) outer space environment (Comedy of the Commons/Anti-commons Scenario (Erlank, 2017, 64)—which could result from excluding others from areas that one cannot effectively use for one's own benefit. The second portion of this acceptance of property rights to celestial bodies or resources is that one's ownership (or other property, or property-like right such as possession, concession, etc) is limited in time and automatically ends when either the reason for exerting control over such portion of celestial real estate ceases to exist, or one loses effective (physical) control over such portion of celestial real estate. For example, if one was mining for Helium-3 on a specific site and all the Helium-3 is exhausted from the area, one should not be able to continue excluding others from the area who might want to use the site for other purposes. The same situation will apply if one was to set up a Moon base, stop using such base, and neither have the inclination, plans, nor ability to return to it and start using it again within a reasonable time, then one's property rights (including the right of exclusion) will automatically cease. It is good to see that such a principle is foreshadowed to a certain extent by the Accords.

2.5 Conclusion

This chapter explores the potential impact of the Artemis Accords on outer space law with a focus on outer space property law. It is shown that the Accords, particularly in Sections 10 and 11 dealing with space resources and safety zones, create a potential conflict with the Outer Space Treaty's free access and non-appropriation principles as set out in Article I and II of the OST. While the OST prohibits national appropriation of outer space and celestial bodies, it is surmised that technological advancements will inevitably lead to resource extraction (of space resources), which will necessitate legal adaptation. The various Sections of the Accords were examined and analysed to determine the implications of the Accords for property rights in outer space. With this in mind, a suggested framework based on effective control and a limited duration of claimed (property or exclusionary property-like) rights to space resources and portions of celestial bodies was proposed as a solution to prevent overreach by space actors (private and State) and to ensure sustainable space utilisation and exploration. While the Accords are an inevitable result of the move to more space exploration, utilisation and exploitation, based on an agreed common interpretation of the non-appropriation principle to allow for the extraction and use of space resources, questions remain about the commercialisation aspects of such extraction and whether this will, in fact, be for the benefit of all mankind. Since the Accords serve as a guiding document, containing a number of agreed-upon principles, but acknowledging and requiring Signatories to cooperate in developing numerous new bilateral and other contractual agreements and practices, it is not perfect, and it is not comprehensive. It suffers from the same vagaries as many other internationally negotiated agreements and treaties by leaving room for misinterpretation and self-interested interpretation of certain aspects. I do believe that it will have clear effects on the practice of space operations and that such practice will eventually evolve into private property-oriented exploitation and use of outer space, its resources and the celestial bodies contained therein. Having a common understanding of how to plan for and undertake a new era of space activities is a noble goal.

References

Blount, P. J. (2011). Renovating space: The future of international space law. *Denver Journal of International Law and Policy, 40*(1–3), 515–532.

Erlank, W. (2015a). Finding property in new places – Property in cyber and outer space. *Potchefstroom Electronic Law Journal/Potchefstroomse Elektroniese Regsblad, 18*(5), 1761–1795. https://doi.org/10.4314/pelj.v18i5.17

Erlank, W. (2015b). Rethinking terra nullius and property law in space. *Potchefstroom Electronic Law Journal/Potchefstroomse Elektroniese Regsblad, 18*(7), 2503–2523. https://doi.org/10.4314/pelj.v18i7.03

Erlank, W. (2016a). Classifying objects in space according to South African property law. *Obiter, 37*(1), 87–105.

Erlank, W. (2016b). Property rights in space: Moving the goal posts so the players don't notice. *Potchefstroom Electronic Law Journal/Potchefstroomse Elektroniese Regsblad, 19*, 1–31. https://doi.org/10.17159/1727-3781/2016/v19i0a1505

Erlank, W. (2017). Property and ownership in outer space. In Y. Abul Failat & A. Ferreira-Snyman (Eds.), *Outer space law: Legal policy and practice* (1st ed., pp. 55–94). Globe Law and Business.

Erlank, W. (2023a). Governance, authority, and chain of command in crewed space voyages: Some thoughts. *Air and Space Law, 42*(2), 165–182. https://doi.org/10.54648/aila2023040

Erlank, W. (2023b). Superficies solo cedit and sectional titles in outer space – An exploratory look. In J. G. Horn, J. M. Pienaar, & M. Wiese (Eds.), *Protean scholar: Essays in Honour of Professor Gerrit Pienaar* (pp. 198–210). Juta.

Erlank, W. (2025). Recognition or prohibition of property rights in outer space: an overview and some caveats. In S. Hobe et al. (Eds.), *Liber Amicorum – 100 years of the Institute of Air Law, Space Law and Cyber Law* (forthcoming) Wolters Kluver.

Ferreira-Snyman, A., & Ferreira, G. (2021). Challenges to the prohibition on sovereignty in outer space – A new frontier for space governance. *PER, 24*, 1–50.

Freeland, S. (2013). Outer space and the non-appropriation principle. In J. C. Smith (Ed.), *Property and sovereignty* (pp. 81–98). Ashgate.

Hobe, S. (2021). The artemis accords what they mean for the development of international space law. *Zeitschrift Fur Luft-Und Weltraumrecht, 70*(1), 1–12.

Hohfeld, W. N. (1917). Fundamental legal conceptions as applied in judicial reasoning. *Yale Law Journal, 26*, 710–770.

Lyall, F., & Larsen, P. B. (2018). *Space law – A treatise* (2nd ed.). Routledge.

Lyall, F., & Larsen, P. B. (2025). *Space law – A treatise* (3rd ed.). Routledge.

Martin, A. S., & Freeland, S. (2022). A round trip to the stars?: Considerations for the regulation of space tourism. *Air and Space Law, 47*(2), 261–284. https://doi.org/10.54648/aila2022014

Masson-Zwaan, T., & Sundahl, M. J. (2021). The Lunar legal landscape: Challenges and opportunities. *Air and Space Law, 46*(1), 29–56. https://doi.org/10.54648/aila2021002

NASA. (2024). *Artemis Accords*. https://www.nasa.gov/artemis-accords/

Planetary Society. (2022, September). *Space policy edition: Mike gold on crafting the artemis accords*. Planetary Radio. https://www.planetary.org/planetary-radio/artemis-accords-mike-gold. Accessed 2024, October 26th.

Pop, V. (2010). Planetary resources in the era of commercialisation. In S. B. Bhat (Ed.), *Space law in the era of commercialisation* (1st ed.). Eastern Book Company.

Svec, M. (2022). Outer space, an area recognised as res communis omnium: Limits of national space mining law. *Space Policy, 60*, 101473. https://doi.org/10.1016/j.spacepol.2021.101473

The Artemis Accords. *Principles for cooperation in the civil exploration and use of the Moon, Mars, Comets, And Asteroids for peaceful purposes* (2020).

Tjandra, J. (2021). The fragmentation of property rights in the law of outer space. *Air and Space Law, 46*(3), 373–394. https://doi.org/10.54648/aila2021021

Tronchetti, F., & Liu, H. (2021). The white house executive order on the recovery and use of space resources: Pushing the boundaries of international space law? *Space Policy, 57*, 101448. https://doi.org/10.1016/j.spacepol.2021.101448

US Department of State. (2024). *Artemis Accords*. https://www.state.gov/artemis-accords/. Accessed 2024, October 26th.

Vertadier, H., & Schingler, J. K. (2023). Safety zones for sustainability on the moon: The example of the breaking ground trust. *Air and Space Law, 48*(SpecialIssue), 191–210. https://doi.org/10.54648/AILA2023038

Xu, F., & Su, J. (2022). Towards a legal regime of benefits sharing for space mining: With some experience from the Area. *Resources Policy, 76*(January), 102627. https://doi.org/10.1016/j.resourpol.2022.102627

Van der Merwe, C. G. (1979). *Sakereg* (2nd ed.). Butterworths.

Chapter 3
Mining in the Province of All Mankind: Analysing Legality with Reference to Treaties and Their Interpretation

Pranav Satyanath and G. V. Ashok

Abstract The principle of non-appropriation of territory enshrined in the Treaty on Principles Governing the Activities of States in the Exploration and Use of Outer Space (Outer Space Treaty) and the Agreement Governing the Activities of States on the Moon and Other Celestial Bodies (Moon Agreement), has thus far stood the test of time and received little criticism by states as neither the technology nor the costs made the appropriation of resources in space was feasible. However, the promulgation of the Artemis Accords has renewed concerns about the legality of mining in space. Relatively recent developments in domestic law and international law, such as Japan, India, Luxembourg, and the United States taking proactive measures to enable the mining of resources, along with the promulgation of the Artemis Accords as a new framework for governing the extraction of space resources, have renewed concerns about the legality of mining in space. Furthermore, renewed competition between China, Russia and the United States has rendered the global order more unstable. In such a volatile climate, where technological progress has gone hand in hand with nations' competing political and economic ambitions, there is an urgent need to precisely assess the permissibility of mining celestial bodies in space. In this chapter, the author explores how space mining can be regulated sustainably under international law. The chapter explores the relationship between the principles of non-appropriation and the principle of outer space being a part of the global commons from the perspective of the rules of interpretation of treaties. Based on such conjoint reading of distinct but related sources of applicable international law, the chapter also revisits the meaning of terms such as ownership" with reference to jurisprudence under the law of the sea.

Keywords UNCLOS · Artemis Accords · Province of all Mankind · Common Heritage of all Mankind · Outer Space Treaty · Seabed Authority

---

P. Satyanath (✉)
SkyServe, Bengaluru, Karnataka, India

G. V. Ashok
Factum Law, Bengaluru, Karnataka, India
e-mail: ashok@factumindia.law

S. Bhat B., A. Variath (eds.), *Artemis Accords and Resource Mining in Outer Space*, Issues in Space, https://doi.org/10.1007/978-981-95-0152-6_3

3.1 Introduction

The Artemis Accords, though slated to be a non-binding instrument, states that mining activities, by themselves, do not breach the principle of non-appropriation governing activities in space. In the Indian Space Policy of 2023, specifically the section titled "Non-Governmental Entities", the Indian Government explicitly recognised the right of such non-governmental entities to, inter alia, "own" and "sell" asteroid resources or space resources. With this approach, India joined the ranks of the United States of America and Luxembourg in announcing its intent to enable the mining and exploitation of outer space.

The Norms, Guidelines and Procedures for the Implementation of the Indian Space Policy, 2023 seem to reflect the substance of the language of the Artemis Accords, merely hinting at what is not violative of the non-appropriation principle as opposed to a positive stand on what is permissible. The stand of nations such as India in pursuing ambitions to leverage the potential resource-abundant outer space as a driver of economic ambitions has reignited the debate around space being a global common. Domestically, one school of thought argues that the reference to space being a "province of all mankind" (United Nations, 1967) in the Treaty on Principles Governing the Activities of States in the Exploration and Use of Outer Space, Including the Moon and Other Celestial Bodies and the principle of non-appropriation ("Outer Space Treaty" or "OST"), excludes the legal permissibility of private actors receiving licenses or being entitled to own and mine outer space. The other school of thought argues that the OST merely prohibits the assertion of sovereign claims over outer space but does not necessarily prohibit state or non-state actors from economically exploiting and benefiting from space.

Central to this controversy lies a question around semantics: what does the phrase "province of all mankind" mean, and does the phrase "national appropriation by claim of sovereignty, by means of use or occupation or by any other means" exclude economic exploitation and use of outer space? In this paper, we seek to chart out a path to ascertain the answers to these questions by reference to principles of interpretation of treaties and examine whether or not the Indian Space Policy of 2023 or the Artemis Accords offend or violate international law.

The experience of exploration driven by the pursuit of business and commerce is vastly different depending upon which part of the world one belongs to. For Europe, as an example, the works of explorers like Christopher Colombus, Vasco Da Gama and the like yielded rich returns. For England specifically, the efforts to reach India, driven by a corporation, the English East India Company, proved to be one of its most profitable ventures, resulting in rich dividends across four centuries. However, for countries like India, which were at the destination port of the ventures of Vasco Da Gama and English East India Company, what followed was a prolonged period of exploitation that reduced a region, otherwise abundant, opulent and self-sustaining, to abject poverty. It is for this reason that the developing world or the Global South has traditionally been sceptical of efforts by nations ahead in technology development and economic growth when it comes to the question of ownership of resources beyond the territorial limits of a nation's sovereignty.

This divide became apparent in the discussions, negotiations and diplomatic processes that culminated in the United Nations Law of the Sea (UNCLOS). While global powers like the USA advocated for lesser restrictions on the access to and use and exploitation of the seas, nations like India, considering the historic inequities perpetuated by centuries of colonial rule, argued for the setting up of a New International Economic Order (Borgese, 1977) that would ensure that the exploitation of seas would benefit all of humanity.

The eventual UNCLOS that emerged was seen as an effort to balance the competing interests of the developed nations with those of the developing states. The result was a consistent theme that saved and advocated for the freedom of navigation in the seas but nevertheless allowed nation-states to declare exclusive economic zones of up to 200 nautical miles from the coast, where the concerned nation-state would have the exclusive right to the resources under the seabed. This principle is enshrined in Article 297 of the Constitution of India, which stipulates that all lands, minerals and other things of value underlining the ocean within the territorial waters or the continental shelf or the exclusive economic zone of India shall vest with the Union of India.

At the heart of the controversy around the legality and permissibility of asteroid mining, lies the concern that nations that benefited from historical inequities, with their advanced technological capabilities, are likely to have first movers' advantage in space. The resulting access to resources in space can have a significant disruptive impact on the terrestrial economies of nation-states on Earth (Oni, 2019). There is precedent already on Earth when companies with exclusive access to precious resources have resorted to disruptive market practices to fix prices and abuse their dominant positions (Saunders, 2000).

While concerns around the mode and manner in which space will be exploited "for the benefit of all mankind" are legitimate, international law has not fared well in addressing the systemic inequalities of international politics. Scholars like Dr. Cassandra Steer, have argued that international law has largely been developed by Western powers to assert and sustain their hegemony in the world order by citing the example of how the United Nations, as an institution, hasn't seen reforms to bring it up to speed with a global reality that is vastly different from the one near the end of World War II that created it in the first place (Steer, 2020).

Seen in the light of these dynamics, the announcements by Luxembourg and the United States of America of a regime to license the extraction and utilisation of resources in space are not only unilateral in character but vindicate the concerns of the Global South when it comes to ensuring a democratic order for the use of outer space. However, there remains the question of how relevant these concerns are in ascertaining the legal position of mining in outer space and the ownership over the fruits thereof.

On the other hand, there is no doubt that exploration of space is expensive and requires capital. NASA's OSIRIS-Rex Mission, which aims to retrieve anywhere between a mere 400 grams to 1 kilogram of material from Asteroid Bennu, is expected to take 7 years and cost over USD 1 billion (Yarlagadda, 2022). Elon Musk, in an interview, estimated that to set up a city on Mars, a million tons of equipment is necessary, which could represent costs upward of USD 1000 trillion, which far exceeds the current Gross Domestic Product of the United States of America (Desk, I. T. S, 2024). If space exploration is going to be as expensive, it remains largely out of reach for most nations on Earth. Even for those better equipped economically, the astronomical costs of going into space is a hurdle and can sometimes raise questions based on a cost-benefit analysis. If the inherent economics of space activities themselves act as a barrier of entry, will the slated objective of space being a "province of all mankind" ever be realised, unless the economic dividends of space, such as through extraction and utilisation of resources in space, subsidise the activity?

Even if one were to ignore the economic dynamics of space activities, there is the question of exploration, even for scientific inquiry and research. As we push deeper into the solar system and beyond, ensuring supply chains between Earth and deep space missions, particularly those which are crewed, will be increasingly challenging not just because it is expensive but because it is time-consuming. For the purposes of ensuring the viability of such crewed deep space missions, mining activities to enable the crew to "live off the land", may be critical. Extraction and utilisation of space resources for sustaining human exploration, which falls under the category of space activities, would require authorisation under the scheme of the OST. While the Global South is justified in its concerns about technologically advanced nations exploiting space to further the gaps between the developed, developing and underdeveloped worlds, it is also equally true that humankind's efforts to go deeper into the final frontier depend on making it economically viable and sustainable. Thus, having sight of these complex, competing and often contradictory objectives and dynamics is critical and relevant for the purposes of arriving at an interpretation of the treaty that is consistent with international law.

3.2 'Province of All Mankind' and 'Common Heritage of all Mankind': Reviewing the Landscape of International Law

Article I of the OST reads as follows, "The exploration and use of outer space, including the moon and other celestial bodies, shall be carried out for the benefit and in the interests of all countries, irrespective of their degree of economic or scientific development, and shall be the province of all mankind. Outer space, including the moon and other celestial bodies, shall be free for exploration and use by all States without discrimination of any kind, on a basis of equality and in accordance with international law, and there shall be free access to all areas of celestial bodies." Article II states that "Outer space, including the moon and other celestial bodies, is not subject to national appropriation by claim of sovereignty, by means of use or occupation, or by any other means."

The term "province of all mankind" is not explicitly defined in the language of the OST itself. The term "province" refers to an area. To digress here, the literal text of Article I notwithstanding, the use of the term "mankind" has also been rightly criticised for being outdated and prejudicial to the reality of the significant contribution made by women in advancing science and technology. However, for purposes of this paper, and we do so with some regret, we shall continue to refer to the literal text as it stands in the hope, however, that at some time in the future, the language will stand corrected to reflect a more inclusive approach to international space law. What does "province of all mankind mean"? Does it then exclude certain kinds of space activities that are driven by the pursuit of business interests? Will it preclude ownership over what is extracted in space?

Article I of the OST itself provides some insight on these questions by stipulating that the exploration and use of outer space shall be carried out for the benefit of and in the interest of all countries. Thus, a bare reading of Article I of the OST seems to suggest that the province of all mankind principle, all nation-states have the right of exploration and use of space, and they shall endeavour to undertake their activities in space for the benefit of all of humanity.

However, Article II of the OST qualifies the text of Article I by suggesting that any use of outer space that has the effect of asserting a sovereign claim is prohibited. The 1979 Agreement Governing the Activities of States on the Moon and Other Celestial Bodies (hereinafter referred to as the Moon Agreement) stipulates that the moon and its natural resources are the common heritage of all mankind (United Nations, 1979).

Those critical of the relevant provisions of the Artemis Accords and the Indian Space Policy, 2023 have argued that a combined reading of Articles I and II of the OST and also the Common Heritage of all Mankind principle of the Moon Agreement still renders any efforts to permit or license asteroid mining as inconsistent with or violative of international law.

However, it appears that this criticism of the provisions of the Artemis Accords and the Indian Space Policy, 2023, appears to adopt a textual or a teleological approach to understanding the provisions of the OST, which begs the question as to whether or not the principles of interpretation of treaties permit such an approach. Before adverting to those principles, to examine what, if any, are the implications of the phrase "Province of All Mankind" and the provision dealing with non-appropriation, it might be useful to refer to a similar phrase appearing in the UNCLOS, that is, the common heritage of all mankind and the relevant provisions reads "Desiring by this Convention to develop the principles embodied in resolution 2749 (XXV) of 17 December 1970 in which the General Assembly of the United Nations solemnly declared inter alia that the area of the seabed and ocean floor and the subsoil thereof, beyond the limits of national jurisdiction, as well as its resources, are the common heritage of mankind, the exploration and exploitation of which shall be carried out for the benefit of mankind as a whole, irrespective of the geographical location of States."

Article 125 states, "Land-locked States shall have the right of access to and from the sea for the purpose of exercising the rights provided for in this Convention including those relating to the freedom of the high seas and the common heritage of mankind. To this end, land-locked states shall enjoy freedom of transit through the territory of transit states by all means of transport." Article 139 lays down that "the Area and its resources are the common heritage of mankind." Article 150 states that "Activities in the Area shall, as specifically provided for in this Part, be carried out in such a manner as to foster healthy development of the world economy and balanced growth of international trade, and to promote international cooperation for the overall development of all countries, especially developing States, and with a view to ensuring: (i) the development of the common heritage for the benefit of mankind as a whole."

A reading of the relevant clauses of the UNCLOS reproduced above suggests that there are some themes in common between the "common heritage" principles set out in the Law of the Sea with the "Province of All Mankind" principle in the OST. These common themes include the emphasis on the freedom of access and navigation. While Article I of the OST refers to the right of use for the benefit of all countries, Article 155 of the UNCLOS refers to the right of exploitation, albeit for the benefit of all countries.[1]

Where the UNCLOS differs from the OST is the fact that administrative and procedural mechanisms to regulate and administer commercial exploitation of the international seabed exist within the framework of the UNCLOS, which has culminated in the setting up of the International Seabed Authority. Exploration and exploitation of the seabed is permitted, subject to a contract with the International Seabed Authority. However, the competing interests and concerns of the developed and developing countries in the mode and manner of functioning of the International Seabed Authority have raised concerns about its functioning and efficacy.

This is seen in the controversy or the lack of consensus on how the UNCLOS can lead to the development of a New International Economic Order and the best means to achieve it (Borgese, 1977). It appears that a fully developed administrative mechanism, without international consensus on how it must function for the exploitation of the international seabed, did result in some disinterest towards exploiting the international seabed for minerals, with one article highlighting how interest in the domain was put on hold for decades (United Nations, 2017).

[1] See Article 155: UNCLOS, The Review Conference. "The Review Conference shall ensure the maintenance of the principle of the common heritage of mankind, the international regime designed to ensure equitable exploitation of the resources of the Area for the benefit of all countries, especially the developing States, and an Authority to organize, conduct and control activities in the Area. It shall also ensure the maintenance of the principles laid down in this Part with regard to the exclusion of claims or exercise of sovereignty over any part of the Area, the rights of States and their general conduct in relation to the Area, and their participation in activities in the Area in conformity with this Convention, the prevention of monopolization of activities in the Area, the use of the Area exclusively for peaceful purposes, economic aspects of activities in the Area, marine scientific research, transfer of technology, protection of the marine environment, protection of human life, rights of coastal States, the legal status of the waters superjacent to the Area and that of the air space above those waters and accommodation between activities in the Area and other activities in the marine environment."

What makes the International Seabed Authority interesting is in assessing the behaviour of nation-states that are not signatories to the UNCLOS, specifically the United States of America. Though the United States of America's stand on the UNCLOS is well known, through the National Oceanic and Atmospheric Administration (NOAA), it has issued licenses to undertake deep seabed mining beyond the territorial waters of the United States of America, going as far back as 1984 (U.S. Congressional Research Services, 2024). Since the UNCLOS came into force, two of the licenses issued by the NOAA have survived and remain in force till 2027, though the ISA designs an area falling within the scope of one of the licenses as an "Area of Particular Environmental Interest". Admittedly not subject to the provisions of the UNCLOS, the USA has not shied away from venturing into the international seabed and staking claim to its dividends.

The analysis of the law of the sea suggests that the common heritage of mankind principle has never been interpreted as prohibiting the ownership over the resources extracted in the area beyond the national sovereignty and jurisdiction of nation-states. On the contrary, a careful reading of the UNCLOS, especially with reference to the provisions dealing with Exclusive Economic Zones and the international seabed, reveals consensus amongst a significant number of nation-states that while sovereignty over regions beyond the territorial waters is not permitted, the same does not necessarily prohibit ownership over resources extracted during mining activities.

The distinction between a nation's legislative jurisdiction, a byproduct of its sovereignty and proprietary rights is a significant feature of the law of the sea. In a catena of judgements of Indian courts as well as English Courts, this was discussed elaborately and the general conclusion has been that even in the absence of absolute sovereign rights, a nation could still take a claim to the benefits derived from the seabed (Datar, 2024). Therefore, mere assertion of ownership over mined or extracted resources would not, under the law of the sea, qualify as an assertion of sovereignty.

Adverting to Articles I and II of the OST as also the CHM principle in the Moon Treaty, the text of the treaties themselves suggest that while the prohibition is against the assertion of sovereignty over outer space, proprietary rights themselves are not prohibited. In practice, a near-earth example is that of orbital slots and spectrum. Under the aegis of the International Telecommunication Union, countries successfully negotiate and ultimately secure limited rights over the use of orbital slots as well as frequencies. In fact, warehousing of orbital slots is not uncommon Centre for International Law, NUS, 2023), and scholarly analysis of the practice, though critical, still alludes to there being no prohibition under the letter of existing sources of international space law (Thompson, 1996). Furthermore, the effect of large constellations being permitted suggests that though *de jure* colonisation of space is prohibited, in practice, advanced space-faring nations often do so (Jah, 2023).

Even after the examination of the landscape across space law and the law of the sea, unfortunately, we are still not closer to the answer to the question we fundamentally seek to answer viz., are the provisions of the Artemis Accords and the Indian Space Policy, 2023 dealing with recognition of the right to extract and utilise resources inconsistent with the OST? In other words, the meaning of space being a province of all mankind or the moon being the subject matter of the common heritage of all mankind remains unclear. The vagueness of these phrases is symptomatic of the said treaties and agreements being the product of political compromises between nations and, thus, reflective of more disagreement than agreement on key areas of interest and controversy (Allott, 1999).[2] How, then, are we to interpret these terms?

3.3 Principles of Interpretation of Treaties and the Outer Space Treaty

The principles of interpretation of treaties are complex and layered. Our efforts to understand these principles are rendered somewhat easier thanks to the Vienna Convention on the Law of Treaties ("VCLT"). Though the VCLT may not provide all the answers to the questions around interpreting the terms of a treaty, it nevertheless is generally applicable. The International Court of Justice (ICJ, 2002), along with other international forums such as the Appellate Body of the World Trade Organisation's Dispute Resolution System, have relied upon and endorsed the VCLT principles concerning the interpretation of treaties as customary international law (Canada v Japan, 1996).

A review of the provisions of the VCLT outlining the interpretation of treaties reveals firstly that the text of a treaty merely is the starting point of the interpretation, and while it is regarded as an authentic expression of the intent of the parties, it is by means no means conclusive. In fact, in addition to the text of the treaty, preparatory materials, records of negotiations, ancillary instruments, interpretative declarations, reservations, and amendments are all relevant to interpreting treaties (Gardiner, 2015). Between these materials relevant to interpretation (hereinafter referred to as "sources of interpretation"), from the text of the treaty to amendments, there is no particular hierarchy that is prescribed as to which source of interpretation is to be afforded primacy (Gardiner, 2015). This has even led some to conclude that a treaty could have not one single autonomous interpretation but many (UK Parliament, 2000).

Furthermore, while the text of a treaty is the starting point of the exercise of interpretation, the ordinary and plain meaning is not always helpful. In the case concerning *Kasikili/Sedudu Island (Botswana/Namibia)* (ICJ, 1999), the International Court of Justice highlighted the limited utility of the "ordinary meaning" of words by highlighting that treaty language sometimes uses terms that do not have a universally accepted "ordinary meaning" and thus the goal of interpretation

[2] Philip Allott's definition of a treaty as 'a disagreement reduced to writing' reveals the nub of the difficulty, containing the implicit warning that one should expect preparatory work to reveal a good deal of divergence and disagreement.

is not to find the "mythical meaning" of such terms. To address the challenges associated with the textual approach, references can instead be also made to the context and objectives attached to the treaty (ICJ, 1999).

For those of us who are trained in English common law, these principles of interpretation of treaties are somewhat in contrast from the principles governing the interpretation of contracts where much emphasis is placed on the language of the contract, with the background negotiations being largely irrelevant and inadmissible for purposes of interpreting them. Unlike the interpretation of contracts between private actors where the objective is to find meaning, the interpretation of treaties cautions against drawing inevitable meanings from the words in a text and against the assumption that the original authors contemplated every future contingency while freezing the language of the treaty. On the contrary, interpretation of a treaty involves giving meaning to the provisions of the treaty, as opposed to finding one (*El Salvador/Honduras: Nicaragua intervening*).

Thus, in analysing the provisions of the OST and the Moon Agreement that define space as a province of all mankind, the moon as the common heritage of all mankind and the principles of non-appropriation, our inquiry places upon us the significantly tall task of giving meaning to those phrases as opposed to ascertaining and finding them. In undertaking this endeavour, the VCLT requires us to refrain from adopting dogmatic textual positivism or an approach that is overly textual or teleological.

With this background, the first question that arises is if the meaning of the terms found in the OST and the Moon Agreement, i.e., "province of all mankind" and the "common heritage of all mankind", is clear or ambiguous. Within the language of Articles I and II of the OST itself, it becomes clear that any use of space that doesn't have the effect of asserting a sovereign claim is technically permitted. Furthermore, even the Moon Agreement doesn't equate the CHM principle to an outright prohibition of ownership over mined resources, instead, it merely suggests that such legality is attached to the coming into force of an international mechanism for administering such activities.

In *Hosaka v. United Airlines*, while examining the provisions of the Warsaw Convention, the United States Court of Appeal held that a text may be considered ambiguous where provisions read together leave open different possible interpretations. Regardless of where one stands on the provisions of the Artemis Accords and the Indian Space Policy, 2023, on the topic of asteroid mining, one must acknowledge that there is room for differing interpretations. In the wake of such ambiguity, one must necessarily turn the other sources of interpretation. It is in this context that we argue that the Artemis Accords, far from being violative of the OST, is merely an effort by the signatories to give meaning to terms and phrases that, by themselves, are not well-defined.

One approach to understanding the role of Artemis Accords in interpreting the OST is the concept of subsequent practice. Article 31 (3) (b) of the VCLT suggests that the subsequent practice of states remains an authentic aid for the interpretation of the treaty. However, provisions dealing with space resources within the Artemis Accords, though they may qualify as a common understanding, still appear to lack clarity and specificity in order to qualify as subsequent state practice under Article 31 (3) (b) of the VCLT (Deplano, 2021). Thus, the concept of subsequent state practice may still not be a useful tool to address the question of what, if any, is the meaning of the terms under reference from the OST.

On the other hand, the concept of "interpretative declaration" as a source of interpretation may be a more useful aid. An interpretive declaration does not modify treaty obligations but only specifies or clarifies the meaning or scope of the terms of the treaty. In the Advisory Opinion rendered in the case of *International Status of South West Africa* (ICJ, 1950), the International Court of Justice held as follows "Interpretations placed upon legal instruments by the parties to them, though not conclusive as to their meaning, have considerable probative value when they contain recognition by a party of its own obligations under an instrument. In this case, the declarations of the Union of South Africa support the conclusions already reached by the Court." The ICJ in yet another case of *the Conditions of Admission to Membership* (Advisory Opinion) [1948] ICJ Reports 63; Ambatielos (UK v Greece) (Preliminary Objection) (ICJ, 1953), described interpretative declarations as being akin to the interpretation clause of a treaty and thus an integral part thereof.

In the context of a multilateral treaty such as the OST, even if one were to assume that the Artemis Accords qualify as an Interpretative Declaration, it is still undoubtedly signed only by a handful of the signatories of the OST. Though this may not be sufficient to give meaning to the terms found in the provisions of Articles I and II of the OST, it still remains a relevant consideration for interpreting the said provisions (Gardiner, 2015).

By itself, the Artemis Accords as an Interpretative Declaration may be seen with some scepticism to arrive at the meaning of the phrase "province of all mankind" in Article I of the OST, but when seen in conjunction with the preparatory materials attached to the OST, we make some progress in the direction of determining the answer to the questions around asteroid mining and the "province of all mankind" principle.

On a perusal of the proposals of the United States of America [*A/AC.105/C.2/L.12*] (UNOOSA, 1966a) and the USSR concerning the text of the OST, it is seen that while the former had actively advocated proprietary rights as a means of protecting investments in space activities, the latter had also promoted free use, access and exploration free of discrimination. Notably, none of the parties had rendered the right of free access, exploration and use subject to the creation of an international administrative mechanism, such as the International Seabed Authority under the UNCLOS.

At this juncture, as the VCLT requires, the context and objective of the OST becomes important. The preamble of the OST only suggests an emphasis on benefit sharing from space activities but does not indicate any intent to negate proprietary rights in space. Furthermore, the emphasis on the right of free access, use and exploration to all, free of discrimination, indicates that the shared expectations of the parties were to mitigate or remedy any barriers of entry that make exploration difficult. There is simply no reason to assume that economic barriers of entry were not meant to be remedied through proprietary rights within the text of the OST itself. If anything, the objective of the treaty appears to further access into outer space, which is not negated by the recognition of proprietary rights.

These references to the sources of interpretation surrounding the OST seem to fortify the contention that the right of use for purposes of extraction and utilisation of resources is not dependent on an international regime for administering the right. On the contrary, if such an absolute encumbrance was intended for the right of extraction and utilisation of resources in space, the sources of interpretation don't evidence it. Furthermore, if the mere absence of an international regime or consensus for administering proprietary rights is sine qua non for recognition of proprietary rights, then the result is essentially that of nations being limited in accessing deeper corners of space which offends the underlying purpose of Article I, namely, the right of free access and use.

3.4 Conclusion

The OST, or the Moon Agreement, was conceived of and came into being at a time when humankind's capabilities and interests in space were in their infancy. Having taken birth in the time before significant technological advancement in space, it is inconceivable that the original authors and signatories had contemplated a scenario where the interest in space would widen to a larger community of non-state actors. As more and more actors see interest in space and seek to navigate the final frontier, a globalised economy also has unique opportunities to address disparities in technologies and capabilities through the device of economics. Central to ensuring non-discriminatory access to space is to provide nation-states with the option of choosing political and economic models that would best suit their context and circumstances. Intrinsic to this thought is the idea that if a nation were to allow explorers to benefit from the exploration, the flow of investment and research and development aimed at the advancement of technology would follow, resulting in inequities being addressed through a diversity of regulatory interventions.

It is in the light of this background that we remember that, ultimately, interpretation of the OST is about giving meaning to the provisions contained in it and not finding one. With the developments in the jurisprudence around evolutionary interpretation, the approach to understanding the meaning of the phrase "province of all mankind" or the non-appropriation principle must be pragmatic and aimed at giving effect to the objectives of the OST in a manner that is most suited for the current realities. Rather than an inquiry into the mythical meaning of these words, the more appropriate approach consistent with the letter and spirit of the VCLT would be to give meaning to them, considering the developments in technology and economic wisdom that are better understood and developed today than in 1967.

References

Allott, P. (1999). The concept of international law. *European Journal of International Law, 10*, 31–50.

Borgese, E. M. (1977). The new international economic order and the law of the sea. *San Diego Law Review, 14*(3), 584. Retrieved from https://digital.sandiego.edu/cgi/viewcontent.cgi?article=1916&context=sdlr

Canada v Japan. (1996). WT/DS8/AB/R.

Centre for International Law, NUS. (2023). Space law: The Kingdom of Tonga and the late-comer problem. *CIL Blog*. Retrieved from https://cil.nus.edu.sg/blogs/space-law-the-kingdom-of-tonga-and-the-late-comer-problem/

Datar, A. (2024, March 9). Legal Notes by Arvind Datar: Do these territorial waters belong to India? *Bar and Bench – Indian Legal News*. Retrieved from https://www.barandbench.com/columns/legal-notes-by-arvind-datar-do-these-territorial-waters-belong-to-india

Deplano, R. (2021). The Artemis Accords: Evolution or revolution in international space law? *International and Comparative Law Quarterly, 70*(3), 799–819. https://doi.org/10.1017/S0020589321000142

Desk, I. T. S. (2024, October 15). Elon Musk's plan to slash Mars colonisation cost from $1,000 trillion to $25 billion. *India Today*. Retrieved from https://www.indiatoday.in/science/story/elon-musks-plan-to-slash-mars-colonisation-cost-from-1000-trillion-to-25-billion-2617332-2024-10-15

Gardiner, R. K. (2015). *Treaty interpretation* (2nd ed.). Oxford University Press.

ICJ. (1950). International Status of South West Africa. *Rep 128.*

ICJ. (1953). Ambatielos (Greece v. United Kingdom). *Rep 10.*

ICJ. (1992). Land, Island and Maritime Frontier Dispute (El Salvador/Honduras: Nicaragua intervening). *Rep 351.*

ICJ. (1999). Kasikili/Sedudu Island (Botswana/Namibia). *Rep 1045.*

ICJ. (2002). Sovereignty over Pulau Ligitan and Pulau Sipadan (Indonesia/Malaysia). *Rep 625.*

Jah, M. (2023). *Occupation, even in orbit, is colonialism*. Aerospace America. Retrieved from https://aerospaceamerica.aiaa.org/departments/occupation-even-in-orbit-is-colonialism/

Oni, D. (2019, September 24). The effect of asteroid mining on mining activities in Africa. *Space in Africa*. Retrieved from https://spaceinafrica.com/2019/09/24/the-effect-of-asteroid-mining-on-mining-activities-in-africa/

Saunders, L. (2000). Rich and rare are the gems they war: Holding de beers accountable for trading conflict diamonds. *Fordham International Law Journal, 24*(4), 1402. Retrieved from https://ir.lawnet.fordham.edu/cgi/viewcontent.cgi?article=1769&context=ilj

Steer, C. (2020). Who has the power? A critical perspective on space governance and new entrants to the space sector. *Georgia Journal of International and Comparative Law*. Retrieved, https://digitalcommons.law.uga.edu/cgi/viewcontent.cgi?article=2492&context=gjicl

Thompson, J. C. (1996). Space for rent: The International Telecommunications Union, space law, and orbit/spectrum leasing. *Journal of Air Law and Commerce, 62*(1), 10.

UK Parliament. (2000). R v Secretary of State for the Home Department, ex parte Adan [2001] UKHL 2 AC 477.

United Nations Office for Outer Space Affairs (UNOOSA). (1966a). (n.d.-a). Draft treaty on principles governing the activities of states in the exploration and use of outer space, including the moon and other celestial bodies: Article I. *(A/AC.105/C.2/L.12)*. Retrieved from https://www.unoosa.org/pdf/limited/c2/AC105_C2_L012E.pdf

United Nations Office for Outer Space Affairs (UNOOSA). (1966b). (n.d.-b). Draft treaty on principles governing the activities of states in the exploration and use of outer space, including the moon and other celestial bodies: Article I. *(A/AC.105/C.2/L.13)*. Retrieved from https://www.unoosa.org/pdf/limited/c2/AC105_C2_L013E.pdf

United Nations. (1967). *Treaty on principles governing the activities of states in the exploration and use of outer space, including the moon and other celestial bodies*. Retrieved from https://www.unoosa.org/oosa/en/ourwork/spacelaw/treaties/outerspacetreaty.html

United Nations. (1979). *Agreement governing the activities of states on the moon and other celestial bodies*. Retrieved from https://www.unoosa.org/oosa/en/ourwork/spacelaw/treaties/intromoon-agreement.html

United Nations. (2017). *The international seabed authority and deep seabed mining*. Retrieved from https://www.un.org/en/chronicle/article/international-seabed-authority-and-deep-seabed-mining

U.S. Congressional Research Service. (2024). *U.S. interest in seabed mining in areas beyond national jurisdiction: Brief background and recent developments*. Retrieved from https://crsreports.congress.gov/product/pdf/IF/IF12608#:~:text=The%20United%20States%20participates%20as,any%20country%20party%20to%20UNCLOS

Yarlagadda, S. (2022, April 8). *Economics of the Stars: The future of asteroid mining and the global economy*. Harvard International Review. Retrieved from https://hir.harvard.edu/economics-of-the-stars/

Chapter 4
The Artemis Accords: An Attempt to Internationally Legalise Outer Space Resource Mining and Ownership?

Tanushri Joshi

Abstract The current international space law framework does not provide clarity on the legality of outer space resource mining, its utilisation and ownership. However, over the past few years, there has been a growing interest between both the Government and private industry to explore this new avenue for Research and Development and commercial purposes. In order to address and curtail the effect of international ambiguity concerning this subject matter, some nations have implemented national legislations permitting outer space resource mining and related aspects. But does national legislation override international treaties and principles? Owning to the economic, societal and legal implications, this paper analyses the legality of outer space resource mining under the current framework of international space law, various national legislations vis-a-vis the international maritime law, and examines whether the Artemis Accords, disguised as soft law, is a step towards internationally validating and reinforcing the space resource extraction, utilisation and addresses its implications.

Keywords NASA · Res communis · Outer Space Treaty · National Laws · Maritime Law · Soft laws

4.1 Introduction

The Outer Space, which is a province of all mankind, is a highly strategic domain influenced by nationalism, protection of one's sovereignty, geopolitics and global cooperation. It is interesting to note that this industry is highly competitive and yet very cooperative at the same time. Accelerating innovation globally, this industry now stands as a pillar of the economy, contributing to international and national socio-economic growth. The moon offers valuable resources, including rare metals,

---

T. Joshi (✉)
Independent Space Lawyer and Business Development Consultant, Mumbai, India
e-mail: tanushri.joshi@outlook.com

S. Bhat B., A. Variath (eds.), *Artemis Accords and Resource Mining in Outer Space*, Issues in Space, https://doi.org/10.1007/978-981-95-0152-6_4

water and Helium-3 (Schmitt, 2007). As outer space, particularly the moon, provides for untapped commercial opportunities, deep space missions, including lunar missions, are now becoming a primary focus of space exploration and further research and development. States today understand the economic importance of outer space resources and already have enabled private organizations to research ways to extract and utilise these resources. Companies like Intuitive Machines, Blue Origin, Astrobotic Technologies, ispace, Planetary Resources, General Motors and many more are focused on developing technologies for lunar missions. In fact, the Agreement Governing the Activities of States on the Moon and Other Celestial Bodies [Moon Agreement] also recognises the benefits derived from the extraction of resources of the moon and other celestial bodies. In 2021, a report evaluating the lunar market by PricewaterhouseCoopers predicted the valuation of the lunar market of €142 billion by 2040 and identified resource mining and utilisation as one of the three ways to measure this output (Luigi & Yann, 2021).

While the Treaty on Principles Governing the Activities of States in the Exploration and Use of Outer Space, including the Moon and Other Celestial Bodies [Outer Space Treaty] permits the use of outer space, including resources on the moon and other celestial bodies, it does not mention of commercial use or utilization of these resources. Commercialisation is linked to trade, i.e., buying or selling goods or services. Generally, for one to have the right to sell, one needs to have ownership or some rights over the goods or subject of trade. But is the extraction of space resources for commercial purposes permitted? Is the appropriation and commercialisation of these extracted space resources permitted? The growing trend in permitting lunar resource extraction and ownership over such extracted resources through the creation of national laws and policies poses a question on the legality of it from an international perspective. In addition to such national laws, what role do the Artemis Accords play in international governance? This article addresses the legality of outer space resource mining and ownership of such extracted resources by examining the treaties, policies and principles surrounding them. At first, it provides more information on the Artemis Accords, then evaluates the current international law and the recently introduced national laws and policies addressing space resource utilisation. As space law is often linked to maritime law, it then analyses whether principles governing marine resource extraction can be applied to outer space resource extraction as well. It then highlights the significance of the Artemis Accords and how its nature of "soft law" has the potential to further lead to the formulation of a binding customary international law and serve as a new mechanism for the creation of international law.

4.2 What Are the Artemis Accords?

The Artemis Accords are a non-binding set of principles introduced by the National Aeronautics and Space Administration (NASA) in coordination with the United States Department of State in 2020. Referred to as a political commitment in the form of a bilateral agreement, the purpose of the Artemis Accords is to guide civil space exploration and use by establishing mutually beneficial practices in the exploration and use of outer space (NASA, 2020). These principles also lay practices focusing on carrying out activities in relation to the Artemis Program. This bilateral agreement, signed at a national and not at an organisational level, aims at implementing the provisions of the Outer Space Treaty and also introduces new concepts concerning lunar exploration and use. Although this United States-led initiative received a lot of criticism, it has 53 signatories as of January 2025. The increasing number of signatories to the Accords reflect the influence it has on space exploration and geopolitics. As the Accords do not create any legally binding obligation, they are not constituted as "hard law" and are not eligible for registration under Article 102 of the Charter of the United Nations.

But if the Artemis Accords are simply a political commitment that cannot be constituted as a treaty, why is the adoption of such a non-binding set of principles a concern? Does it significantly impact the international governance and treaty-making process?

4.3 Why Are the Artemis Accords Stirring Up the Debate?

Prima facie, the Artemis Accords appear to only further the international cooperation and other provisions of the Outer Space Treaty. Supporting provisions of the Outer Space Treaty, the Artemis Accords establish a set of best practices and norms of responsible behavior for civil space exploration and use namely: Activities to be carried out for peaceful purposes, transparency, interoperability, assistance to personnel in outer space, registration of space objects, release of scientific data, preservation of outer space heritage, utilisation of space resources, deconfliction of space activities and sustainable use of outer space (NASA, 2020). However, certain provisions of the Artemis Accords, namely Section 10—Space Resources and Section 11—Deconfliction of Space Activities, are said to drift away from the principles of the Outer Space Treaty. Section 10 recognises the importance of utilisation of space resources and how it can benefit the humankind. However, it further states that such extraction does not constitute national appropriation under Article 11 of the Outer Space Treaty (NASA, 2020). Section 11 of the Artemis Accords (NASA, 2020) tries to implement the obligation of states under the Outer Space Treaty by asking states to notify and coordinate with other states concerning the operations undertaken and further speaks about creation of safety zones to avoid any harmful interference (NASA, 2020). Some view this provision to eventually permit appropriation under the garb of creation of "safety zone" and curb the freedom of exploration and the concept of *res communis*. But before we further analyse these provisions, let us understand the current provisions surrounding freedom of exploration, use, extraction of space resources.

4.4 The Current International Scenario

The current international framework governing the activities in outer space consists of five main treaties, of which the Outer Space Treaty and Moon Agreement are relevant to our current discussion. The Outer Space Treaty, which entered into force on October 10, 1967, is known as the Magna Carta of Space Law (Maurice, 2004). This treaty, formed to protect outer space from the effects of two world wars, the Cold War, and the arms race, and to cap the ambitions of establishing one's sovereignty in outer space, propounds important principles concerning the outer space exploration and utilisation both by the government and the private entities. Although the Outer Space Treaty does not specifically address the legality of outer space mining, it lays down important principles surrounding this topic through the lens of "res communis" concept:

1. Article 1 (Outer Space Treaty, 1967) states that outer space is the province of all mankind and is free for exploration and use by all states. There is free access to all areas of celestial bodies;
2. Article 2 (Outer Space Treaty, 1967) prohibits the national appropriation of the moon and other celestial bodies by claims of sovereignty through either use, occupation or any other means
3. Article 3 (Outer Space Treaty, 1967) states that all activities concerning the exploration and use of outer space, including the moon and the other celestial bodies, shall be in accordance with international law, including the Charter of the United Nations.

The Outer Space Treaty has been operative for more than five decades today and has been largely successful in preventing any warfare in outer space until now. It has also fostered the development of outer space as a province of mankind. However, it lacks precise definitions of some crucial terms and fails to offer specific regulatory guidance. This treaty prohibiting national appropriation of outer space, including the moon and other celestial bodies, does not shed light on how "use", "occupation" or "any other means" may result in appropriation. But at the same time, Article 1 of the Outer Space Treaty states that the outer space, including the moon and other celestial bodies, are free for exploration and use by all states. This vagueness can lead to the fragmentation of its interpretation as the activities and space law governance evolve (Brian, 2014). Thus, the Outer Space Treaty, lacking any concrete stance on what kind of "usage" leads to appropriation, whether extraction comes under the purview of permitted use of the outer space or not, sparks the legal uncertainty around the space resource mining.

The Moon Agreement is an expansion of the principles laid down in the Outer Space Treaty. However, unlike the Outer Space Treaty, the Moon Agreement specifically addresses some ambiguity and prohibits claiming or appropriating any resources. Reflecting the principles of the Outer Space Treaty, the Moon Agreement states that the moon is not subject to national appropriation by claim of sovereignty, or by means of use or occupation, or by any other means and regards the moon and its natural resources as the common heritage of mankind. It further clarifies that neither the surface of the moon, nor the subsurface of the moon nor any natural resources in place can become property of any State, international intergovernmental or non-governmental organisation, national organisation or non-governmental entity or of any natural person. It also prohibits any claim or right of ownership over the surface or subsurface of the moon by placing of any personnel, space vehicles, equipment, facilities, etc., either on or below the surface of the moon (Moon Agreement, 1979). Unfortunately, the Moon Agreement failed to get international support not due to any ambiguity but for the reason that it provides much more clarity on the utilisation of space resources and prohibits ownership of space resources by anyone. Ratified by a handful of states, this Moon Agreement has seen a couple of withdrawals lately. Since the Moon Agreement is not widely accepted, its value as a source of interpretation of certain principles governing outer space activities is often questioned. Rather, its rejection by many state parties to the Outer Space Treaty may indicate the alternative preference of governance surrounding lunar activities.

Interestingly, the Moon Agreement also fails to provide definition of the term "use" of the moon and its natural resources. Some prefer to interpret the term "natural resources in place" differently and often use this as a tool to rationalise that the resources once extracted from the surface can be utilised, and ownership over such extracted resources can be claimed.

Some construe that the Outer Space Treaty may permit for outer space mining as compared to the Moon Agreement, which specifically prohibits ownership rights over any natural lunar resource. In cases where it becomes difficult to interpret a treaty, the Vienna Convention on Law of Treaties (VCLT) sets out ways to interpret the treaty. At first, according to Article 31 of VCLT, a treaty has to be interpreted in good faith by giving ordinary meaning to the terms and has to be read in the context of its object and purpose (VCLT, 1969). Besides, as per Article 32 (VCLT, 1969), the preparatory work of the treaty and the circumstances of its conclusion can be referred to as a supplementary means of interpretation. Thus, a collective reading of both the Outer Space Treaty, the Moon Agreement, the primary reasons for the creation of the Outer Space Treaty and other international space law treaties points out that appropriation through any form or means is not permitted. Yet, the lack of explicit prohibition in the Outer Space Treaty and lack of signatories to the Moon Agreement add to the ambiguity surrounding the extraction of natural resources of the moon and its ownership.

4.5 No Race Between the Space Laws: International Law at Standstill While National Laws Pace Up

If we take a look at the way space law has been, the space activities led to the formation of space law. Although space law was proactive during its formative years, it has not paced up to the technological innovation we see today. The increase in private participation and ever-changing geopolitical dynamics has favoured many states to build policies and laws comforting the domestic environment within such states. This is for the following reasons:

1. Space as an industry is evolving at a higher pace and will continue to do so.
2. It is a highly capital-intensive market, which translates to a high-risk market as well.
3. The investors would like to have certainty, including legal certainty, concerning activities carried on by companies they have invested in or wish to invest in.

The United States was the first nation in the world to have a national law permitting outer space resource extraction and its use. Further, by authorising Moon Express Inc. to travel beyond Earth's orbit and to land and explore the Moon's surface changed the perception towards the private lunar exploration and its related licensing. This authorisation by the Federal Aviation Authority (FAA) reflected the government's understanding of the importance of commercial lunar exploration and the need to discover the potential of lunar resources (OECD, 2016).

The United States legalised space mining by introducing the US Commercial Space Launch Competitiveness Act, 2015. In order to facilitate the commercial exploration and recovery of space resources, Article 51,303 states that "A United States citizen engaged in commercial recovery of an asteroid resource or a space resource under this chapter shall be entitled to any asteroid resource or space resource obtained, including to possess, own, transport, use, and sell the asteroid resource or space resource obtained in accordance with applicable law, including the international obligations of the United States" (US Commercial Space Launch Competitiveness Act, 2015).

The introduction of this law, especially the clear focus on commercialisation, raised concerns around the United States's interpretation of provisions of the Outer Space Treaty. Later, the United States also released an executive order to encourage international support for public and private extraction and use of resources in outer space, negotiate joint statements, bilateral and multilateral arrangements with foreign states regarding safe and sustainable operations for public and private recovery and use of space resources (Executive Order 13914, 2020). It must be pointed out that the United States is not the only nation to do this. Luxembourg became the first European state to adopt regulations permitting space mining. Article one of the Law on the Exploration and use of Space Resources states that outer space resources are capable to be owned (Law of July 20th 2017 on the exploration and use of space resources, 2017). Luxembourg's SpaceResources.lu is an initiative led by the Luxembourg Space Agency with the objective of placing Luxembourg as a pioneer in the exploration and utilisation of space resources (Luxembourg Space Agency, 2022). Consistent with UAE's 2030 Strategy, the United Arab Emirates Federal Law 12 (2019) enables and permits exploration and exploitation of non-living space resources and further allows extraction, recovery, refining, processing, use, ownership, purchase, sale, trading, transportation or storage of such resources. Japan's Space Resources Act (2021) also provides ownership of mined space resources by licensed companies. Furthermore, the Indian Space Policy 2023, encourages non-governmental entities to engage in the commercial recovery of asteroid resources or space resources. It permits such non-governmental entities to possess, own, transport, use, and sell such asteroid resources in accordance with Indian laws and India's international obligations. Furthermore, it states that the Norms, Guidelines and Procedures for Implementation of Indian Space Policy-2023 in respect of Authorization of Space Activities refrains the non-governmental entities from asserting any sovereign claims on the extracted resource but at the same time also recognizes that not all forms of extraction and utilization of resources in space can qualify as appropriation.

Not only in the United States but also in other countries, the space industry has seen immense participation and contribution by the private industry. As many states want to wisely spend their resources on the development of strategic technology and on other research and development, states are supporting the private space industry and private investment in the space industry, including the acceptance of foreign investments. In order to further promote private participation in the space business, many states offer incentives and benefits such as tax waivers or reductions, grants without interest and much more. Recently, under the ease of doing business, many states also offer the establishing of a private sector favourable law and policy. Apart from the favourable internal law and policies, Luxembourg, understanding the strategic and financial importance, offers financial support for organisations wanting to undertake space mining activities.

Furthermore, many have signed bilateral agreements with states that permit space resource extraction and mining. Japan and Luxembourg entered into a five-year cooperation agreement to explore and commercially utilise of space resources (Ministry of Economy-Luxembourg, 2017). Subsequently, many states, including Portugal, UAE (The Luxembourg Government, 2017) and even inter-governmental agencies like the European Space Agency (ESA) entered into a similar cooperation agreement with Luxembourg. In the past, NASA's OSIRIS-Rex spacecraft carried samples from asteroid Bennu, Hayabusa-1 and Hayabusa—2—Japan's spacecrafts extracted samples from two asteroids namely, Itokawa and Ryugu. As of 2023, a private company named ispace launched a robotic arm "HAKUTO-R", which is currently orbiting around the moon (ispace, 2023). China's Chang'e-6 spacecraft successfully brought lunar samples to earth (Ian, 2024). In fact, this was also displayed at the 75th International Astronautical Congress, which took place in Milan, Italy (Deng, 2024).

While there is a growth in the number of states permitting space resource mining and ownership over such extracted resources, there is a need for an active international framework to ensure that such resource extraction is beyond selfish national interest and instead establishes cooperation, provides equal access and promotes sustainable practices concerning these activities.

4.6 Confusion Created by National Laws and International Laws: Is Maritime Law to the Rescue?

The practice of running to international maritime law to interpret and analyse establishing of rights and practices in the exploration and use of outer space is not uncommon. The main reason for this is the common set of problem it needs to address. Like the outer space, the high seas and the sea bed hold a plethora of minerals. Like outer space, international waters are also free to be used and accessed by all states. However, there lies a difference with regards to renewability of resources in these two regions. The argument of "owning the fish and not the sea" cannot be utilised here as fish are living resources that can reproduce, thus making them renewable unlike the outer space resources which are capable of being depleted.

Like the Outer space, even the deep sea-bed area which includes the seabed and ocean floor and subsoil thereof is beyond the limits of national jurisdiction and is open to use by all States. Although sovereign claims over the deep seabed area are prohibited, the United Nations Convention on the Law of the Sea (UNCLOS) allows parties to establish property rights in extracted resources without violating the non-appropriation principle. However, the natural resources of the seabed are governed by a complex legal regime as stated under Article 136 of the (UNCLOS, 1982). Article 137 vests the rights over the deep seabed resources in the International Seabed Authority (ISA) (UNCLOS, 1982). Thus, state parties wanting to undertake mining operations must be authorized by ISA before undertaking any such activity. The same principle is referred to in Article 11 (5) of the Moon Agreement urging states to undertake and establish an international regime and appropriate procedures to govern the exploitation of natural resources of the moon. But the Moon Agreement is not ratified by many States and considering the overall approach by majority of states, there would be no efforts towards implementing this principle mentioned in the Moon Agreement. Thus, considering the vast differences between the outer space and the high seas, we cannot simply apply the same rationale while creating policies or laws governing outer space activities.

4.7 Who Prevails Over Whom: National v. International Law?

Given the rise of national laws, especially governing the outer space resource mining activity, an important question arises. Can national laws prevail over international law? According to Article 26, international treaties bind only states that are parties to it (VCLT, 1969). Once a state is a party to a treaty, it cannot breach such treaty obligations by creating national laws either permitting or prohibiting such act. This principle has been reaffirmed multiple times (Permanent Court of International Justice, 1930). "The characterization of an act of a State as internationally wrongful is governed by international law. Such characterization is not affected by the characterization of the same act as lawful by internal law" (Responsibility of States for Internationally Wrongful Acts, 2001). Is the creation of national laws and policies permitting the use and extraction of outer space resources and claiming ownership over such extracted resources an attempt to override the Outer Space Treaty? The current national laws and policies of various states uphold most principles of Outer Space Treaty. However, they take advantage of the loopholes in the Outer Space Treaty, and provide a different perspective on the interpretation of "use" "extraction of resources". While these national laws do not permit ownership of the land, they add an element of ownership to such extracted resource, which, by strict interpretation of Outer Space Treaty, inherently violates the principle of "outer space is a province of mankind". On one hand we have the national laws permitting ownership over such extracted resource and on other hand we have a set of principles in the form of Artemis Accords, which state that "extraction" of outer space resources does not amount to national appropriation. But as states cannot justify their failure to perform a treaty by creating any internal law, what is the purpose of taking a different stance through these internal laws? The absence of an international law addressing this grey area surrounding resource extraction and its ownership leaves room for further interpretation and sows the seed for formulating interpretations suitable for most states and their long-term activities. The question arises as to what is the importance of these national laws having a different position on ownership of space resources. What is the need of the Artemis Accords? Is there any correlation between these national laws and the Artemis Accords in the long run?

4.8 Why Artemis Accords? What Is the Significance of the Artemis Accords Here?

As international law could not form any consensus on whether space mining is permitted or not, some states undertook legalising of the outer space resource mining upon themselves by drafting national space laws and policies favouring resource extraction. But if an international instrument like Moon Agreement already exists, why not amend it? Or why not create rules or implementation guidelines in furtherance to the Moon Agreement?

If the intention is to provide clarity on the interpretation of an international treaty, why create a set of principles that too in a form of a non-binding bilateral agreement? Is this an intentional step towards something much larger?

As stated earlier, the Artemis Accords are primarily set of norms, principles that uphold the fundamentals stated in the Outer Space Treaty. At the same time, these Accords reiterate the notion as supported by many national laws and policies that the outer space resource extraction and use of such extracted resources does not amount to national appropriation. In times when the legality of space resource mining under international law remains controversial not only among states but among individual experts as well, it is pertinent to have some guiding force to shape the evolution of space activities, space governance and norms. This is especially true in the light of rapid increase surrounding lunar activities.

The multilateral treaty-making process is time-consuming and may not support the current technological advancements. Moreover, if we consider the development of outer space law, the formation of space law in the form of multilateral treaties is inactive and has remained stagnant for years. Further, the failure of the Moon Agreement to get enough signatures created concerns surrounding the current treaty-making process. With the decline in this traditional approach, states prefer "soft law" as an alternative process for reaching an international consensus on laws governing outer space activities (Fabio, 2011).

Soft Law is generally a set of non-binding principles in the form of norms, bilateral or multilateral memorandums or agreements, concluding texts of conferences, recommendations, executive political agreements, guidelines and much more (Shelton, 2008). Soft law "expresses a preference and not an obligation that state should act, or should refrain from acting, in a specified manner." (Gold, 1996). This "expressed preference" for certain behaviour aims to achieve functional cooperation among states to reach international goals (Lichtenstein, 2001).

Soft law aims to create a set of standards and best practices concerning the subject. Furthermore, as these are not binding, states have a preference to enter into these agreements as compared to something that obliges them to do or refrain from doing a particular act, failure of which could result in a breach and attract penalties or sanctions.

4.9 Lifting the Artemis Accord Veil: Relevance of Soft Law in the Interpretative Process

Lifting of corporate veil is a concept in corporate law, where in order to assess the business and determine the actual intention of creation and operation of such corporate entity, authorities look beyond the legal identity of the company, by examining the company and its members who are in control of running such companies. If found guilty, the owners, shareholders, and directors are held liable for such company's actions or inactions.

In this current scenario, it is pertinent for us to look beyond the Artemis Accords as a mere political commitment and understand if there is any particular intent in creating a bilateral agreement instead of moving this as a resolution in the United Nations General Assembly. Although the Artemis Accords were presented as a guide for NASA and its international partners with regards to their lunar missions (NASA, 2020), they hold a much broader and powerful force in the form of a normative instrument.

Along with upholding the principles of the Outer Space Treaty, the overarching goal of the Artemis Accords is to gain consensus that the extraction and utilization of space resources, including any recovery of such space resources, does not amount to national appropriation. To discuss such a topic and gain the assent of the majority of nations would be extremely time-consuming. On the other hand, there is the ever-growing pressure to create and ensure legal stability conducive to encourage investments and ensure access to the outer space (The Economist, 2019).

Keeping this in mind, introducing the Artemis Accords in the form of soft law is a smart move which has the potential to further influence and contribute to international law.

4.9.1 Soft Law as a Step Towards Building Customary International Law

International law is not restricted to merely treaties. Article 38 of the International Court of Justice recognizes the following as sources of International Law:

a. International conventions;
b. International Custom, as evidence of a general practice accepted by law;
c. The general principles of law evidenced by civilized nations and
d. Judicial decisions and teachings of the most highly qualified publicists of various nations

A deep dive into the creation of Artemis Accords is to further support their interpretative notion around space resource extraction and use and establish it as a customary international law. Customary International Law is not written but a set of uncodified practices undertaken and recognised by multiple nations. However, customary international law can be created only if it satisfies two elements: Practice and Opinio Juris (Christopher, 2008).

Soft law presents itself as a great instrument in the creation of practice and opinion juris. As they bring in inclusivity and cooperation at an international scale without an arduous process and generally eliminates the binding structure of an international treaty (Christensen & Johnson, 2020), soft law is a preferred mechanism and has often been used to overcome ambiguities within existing treaties and conventions. In fact, the United Nations Legal Subcommittee developed soft law principles to resolve and address international issues, e.g., use of nuclear power sources in outer space (Jennifer, 2016). Unlike other previous soft law instruments drafted by United Nations Committee on Peaceful Uses of Outer Space and introduced at the United Nations General Assembly, the Artemis Accords are a United States drafted and United States led initiative.

Soft law is referred to as "long established, vital component of the space law regime" (Beard 2017). However, soft laws have the potential to be referred to for interpretation of treaties and further create procedural obligations as well.

The VCLT under Article 31 throws light on the interpretation of treaty. A treaty at first has to be interpreted in good faith and has to be read in the light of its object and purpose (VCLT, 1969). Further, Article 31(3) highlights that subsequent state agreement and subsequent state practice can be referred to interpret a treaty (VCLT, 1969).

In the present scenario, the creation of national space laws and policies by various states, the introduction and signing of the Artemis Accords all can be considered as subsequent state practice in order to interpret the legality of space resource. A closer look at the actions of the United States is enough evidence to support the interpretative initiative of the United States. In addition to the Artemis Program, it first introduced the Commercial Space Launch Competitiveness Act, 2015, then supported and authorised Moon Inc., a private entity to explore the lunar surface, passed an Executive Order to encourage international support for recovery and use of outer space resources, presented the Artemis Accords as a non-binding instrument and then established grants and offers to purchase materials collected from the moon by the commercial operators.

Furthermore, the Artemis Accords do not as such require the signatories to have a national law recognising the interpretation of the Outer Space Treaty. However, it is pertinent to note the passing effect the Artemis Accords create. For example, if a private entity of state A enters into a bilateral agreement with another private entity of state B, which is signatory to the Artemis Accords and/or has national law and policy permitting it to undertake such activity, for private entity of State A to carry out such activities, it needs permission of the Government of state A to do so. This would in turn require State A to introduce a set of policy or law governing such activities or establish a set of unwritten practices. Ultimately, it makes state A to accept and recognize such subsequent practice. And given the collaborative nature of the space industry and rise in privatization globally, no nation would deny any such activity unless it is against their national interest.

Soft laws can become "hard laws" if they are adopted by states in their national laws (*Boyle, 2008*). Presently, national laws permit respective entities to extract and use the outer space resources and also give title over such extracted resources. Moreover, the Artemis Accords interpret the extraction and use of space resources as non-appropriation, a combined reading of the national laws and Artemis Accords reflects as a state practice and pre-existing customary international law. Ultimately this customary international law can be considered as a binding law. Keeping aside the existence of Artemis Accords and other national laws and policies surrounding space resources, if we were to consider amending the Moon Agreement, it would need the consensus of States Parties to do so. However, as the Moon Agreement has not been ratified largely, the amendment of the Moon Agreement will not create any further impact unless it is then ratified by States.

4.10 Conclusion

Given the economic and strategic benefits of valuable minerals and other resources available in outer space, the extraction of outer space resources in the coming years is inevitable. The Outer Space Treaty permits the use of outer space, including its resources, giving clarity on what kind of "use" is permitted and what amounts to "appropriation" will provide legal assurance for national or international organisations and other private organisations to carry out their activities. Currently, important terms like what constitutes a "space object", "appropriation", where does outer space begin, and many other important terms and concepts remain undefined. As the activities towards exploration of outer space have immensely grown globally, the need to determine and define these key terms is of a high priority. The creation of the Artemis Accords, which are considered as a guiding set of principles for lunar activities, cannot be viewed in a negative sense in toto. Such non-binding instruments are a great way to guide the formulation of laws, especially the ones governing ever-evolving technology. With the growing private interest and availability of private funds, legal security and certainty are required to sustain this highly capital-intensive market. While such instruments can be considered as a way to guide towards treaty-making, they shall not dictate the law itself. The Artemis Accords read together with national laws will formulate customary international law, which will serve as a binding international law. Although states have the freedom to become signatories to a particular treaty or not, the approach of interpreting terms of the existing binding treaties by creation of national laws and policies first, instead of discussing at multilateral forums such as the United Nations Office of Outer Space Affairs will question the entire purpose of United Nations in the coming years. States ought to cooperate and coordinate not only on the technical matters and projects but also on the legal affairs in order to prevent the Moon from being an international area of conflict. Can a framework similar to ISA be considered for lunar resources? Or perhaps the manner in which the International Telecommunication Union regulates the radio frequency spectrum, which is also a limited natural space resource, can be referred as well. These questions have to be addressed and discussed at an international forum. Nevertheless, the creation of the Working Group on Legal Aspects of Space Resource Activities is a step towards collectively creating international laws, policies and principles surrounding resource mining.

References

Agreement Governing the Activities of States on the Moon and Other Celestial Bodies, opened on signature on 18 December 1978, 1363 U.N.T.S. 22, 18 I.L.M. 1434 (1979) (entered into force on 11 July 1984) [Moon Agreement].

Beard, J. M. (2017). *Soft Law's Failure on the Horizon: The international code of conduct for outer space activities*. 38 U. Pa. J. Int'l L. 335. https://scholarship.law.upenn.edu/jil/vol38/iss2/2

Boyle, A. E. (1999). Some reflections on the relationship of treaties and soft law. *International and Comparative Law Quarterly, 48*(4), 901–913. https://doi.org/10.1017/S0020589300063739

Brian Israel. (2014). Treaty Stasis. *AJIL Unbound, 108*, 63, 64, 65. https://doi.org/10.1017/S2398772300001860

Christensen, I. A., & Johnson, C. D.. (2020). *Putting the White House executive order on space resources in an international context*. https://www.thespacereview.com/article/3932/1

Christopher, W. (2008). *Sources of international law: An introduction*. https://legal.un.org/avl/pdf/ls/greenwood_outline.pdf

Deng, X. (19 October 2024). *China's Chang'e-6 lunar samples make global debut at IAC in Italy.* Global Times. https://www.globaltimes.cn/page/202410/1321483.shtml

Executive Order 13914. (2020). 85 Fed. Reg. 20,381. https://www.federalregister.gov/documents/2020/04/10/2020-07800/encouraging international-support-for-the-recovery-and-use-of-space-resources

Fabio, T. (2011). Soft law. In C. Brünner & A. Soucek (Eds.), *Outer space in society, politics and law* (p. 619). Springer.

Gold, J. (1996). *Interpretation.* The IMF and International Law.

Ian, S. (25 June, 2024). *China's Chang'e-6 lunar probe returns world's first samples from far side of the moon.* The Guardian. https://www.theguardian.com/world/article/2024/jun/25/chinas-change-6-lunar-probe-returns-world-first-samples-from-far-side-of-the-moon

ispace. (2023). "ispace Announces Results of the "HAKUTO-R" Mission 1 Lunar Landing", press release, 26 May 2023, https://ispace-inc.com/news-en/?p=4691

Jennifer, U. (2016). Soft law: The key to security in a globalized outer space. *Transportation Law Journal.* https://doi.org/10.2139/ssrn.2718328

Law of July 20th 2017 on the exploration and use of space resources. (2017). Retrieved from https://spaceagency.public.lu/en/agency/legalframework/law_space_resources_english_translation.html

Lichtenstein, C. (2001). *"Hard Law v. Soft Law: Unnecessary Dichotomy?"* 35 Int'l Law. 1433.

Luigi, S., & Yann, P.. (2021). *Lunar Market Assessment: market trends and challenges in the development of a lunar economy.* https://www.pwc.com.au/industry/space-industry/lunar-market-assessment-2021.pdf

Luxembourg Space Agency. (2022). *Contribution of the grand duchy of luxembourg on the mandate and purpose of the working group on legal aspects of space resource activities.* https://www.unoosa.org/documents/pdf/copuos/lsc/spaceresources/LSC2023/StatesResponses/Luxembourg_-_20221216_WG_SR_LU_Contribution.pdf

Maurice, N. A. (2004). *The 1967 Outer Space Treaty (1967 OST) as the Magna Carta of contemporary space law: A brief reflection.* https://www.elevenjournals.com/tijdschrift/iisl/2004/3%20A%20General%20Convention%20on%20Space%20Law%3F/IISL_2004_047_003_006.pdf

Ministry of Economy-Luxembourg. (2017). *Luxembourg and Japan agree to cooperate on exploration and commercial utilization of space resources.* https://space-agency.public.lu/dam-assets/press-release/2017/2017-11-29-press-release-mou-japan-space.pdf

National Aeronautics and Space Administration. (2020). *The Artemis Accords: Principles for a safe, peaceful, and prosperous future.* Retrieved from: https://www.nasa.gov/specials/artemis-accords/index.html

OECD. (2016). *Space and innovation.* OECD Publishing, Paris. https://doi.org/10.1787/9789264264014-en

Permanent Court of International Justice. (1930). *Greco-Bulgarian "Communities"* (Advisory Opinion) (Series B, No. 17, p. 32).

Responsibility of States for Internationally Wrongful Acts, adopted on 10 August 2001. UN Doc. A/RES/56/83 (2001). 53 UN GAOR Supp. (No. 10) at 43, Supp. (No. 10) A/56/10 (IV.E.1). https://legal.un.org/ilc/texts/instruments/english/draft_articles/9_6_2001.pdf

Schmitt, H. (2007). *Return to the Moon: Exploration, enterprise, and energy in the human settlement of space.* Springer. https://books.google.co.uk/books?id=IerrQGC6S2YC

Shelton, D. L. (2008). *Soft law in handbook of international law.* https://scholarship.law.gwu.edu/cgi/viewcontent.cgi?params=/context/faculty_publications/article/2048/&path_info=Soft_Law.pdf

The Economist. (2019). *Space law is inadequate for the boom in human activity there.* https://www.economist.com/international/2019/07/18/space-law-is-inadequate-for-the-boom-in-human-activity-there

The Luxembourg Government. (2017). *Luxembourg and the United Arab Emirates to cooperate on space activities with particular focus on the exploration and use of space resources.* Retrieved from https://gouvernement.lu/fr/actualites/toutes_actualites.gouvernement2024%2

Bfr%2Bactualites%2Btoutes_actualites%2Bcommuniques%2B2017%2B10-octobre%2B10-uae-space.html

The United Nations Convention on the Law of the Sea, opened for signature 10 December 1982, 1833 UNTS 397, 21 ILM 1261 (entered into force on 16 November 1994).

Treaty on Principles Governing the Activities of States in the Exploration and Use of Outer Space, including the Moon and Other Celestial Bodies, opened for signature 27 January 1967, 610 UNTS 205, 18 UST 2410, TIAS No 6347, 6 ILM 386 (entered into force on 10 October 1967) [Outer Space Treaty].

US Commercial Space Launch Competitiveness Act. (2015). 89 FR 76714. Retrieved from https://www.federalregister.gov/documents/2024/09/19/2024-20900/us-commercial-space-launch-competitiveness-act-incorporation

Vienna Convention on the Law of Treaties, opened for signature 23 May 1969, 155 UNTS 331, 8 ILM 679 (1969), 63 AJIL 875 (entered into force on 27 January 1980) [VCLT].

Chapter 5
The Interplay Between the Moon Agreement and the Artemis Accords: Vienna Convention on the Law of Treaties to the Rescue

Atul Alexander

Abstract When Sputnik and Explorer 1 Satellites were successfully launched to orbit the Earth, the Soviet Union and the U.S. set their sights on the Moon. The Soviet Union performed the first hard landing with the Luna 2 Spacecraft in 1959, and Neil Armstrong from the U.S. became the first man to set foot on the Moon. Since these unimaginable feats, two other States joined the elite club, i.e., China and India. But none of these States have attempted to exploit the Moon's resources. The Moon Agreement, 1979 tries to balance exploration and exploitation; however, Article 11 proscribes the individual commercialisation and appropriation of the Moon and its natural resources. The U.S., under the Trump administration, drafted the Artemis Accord in 2020, purportedly in line with The Outer Space Treaty 1967. The Accord is a series of bilateral arrangements between the U.S. and other governments to regulate commercial activities in Space, including the Moon. Hence, there is a discernible tension between the Moon Agreement and the Artemis Accords, the latter being pro-commercialization and the former prohibiting commercialisation for individual benefit. These developments affect the State's simultaneous signatories to these instruments. States also appear to have taken a chequered position; Australia considered the Artemis Accords to be in sync with the Moon Agreement, while Saudi withdrew from the Moon Agreement as it is incompatible with the Artemis Accords. In this chapter, the author applies the rules of the Vienna Convention on Law of Treaties (VCLT), 1969, and attempts to decipher and interpret these instruments. Although several scholars have written on the interplay between them, there is a dearth of literature specific to VCLT.

Keywords VCLT · Moon agreement · Special regimes · United Nations · ICJ · WTO

A. Alexander (✉)
The West Bengal National University of Juridical Sciences, Kolkata, West Bengal, India
e-mail: atulalexander100@nujs.edu

S. Bhat B., A. Variath (eds.), *Artemis Accords and Resource Mining in Outer Space*, Issues in Space, https://doi.org/10.1007/978-981-95-0152-6_5

5.1 Introduction

The advancement of science and technology has resulted in States looking at the Moon and other celestial bodies as an alternative to the Earth (Tronchetti, 2009). In this background, NASA is coordinating with other private agencies for the development of manned space flights named the Artemis programme (NASA, 2020). One of the features of the Artemis programme is to construct a permanent settlement on the Moon, which includes a lunar base and Gateway (Malik, 2023). States wanting to be part of the programme are required to comply with the Artemis Accord (Grush, 2020). The Accord was launched courtesy of the American plan to send the first woman and the next man to the Moon by 2024 (ET Bureau, 2020). According to NASA administrator Jim Bridenstine, "Fundamentally, the accords are about avoiding conflict, transparency, public registration, de-conflicting activities." (Foust, 2020) Speaking about the Accord, Frans von der Dunk "drew parallels with the development of international civil aviation regulations, which started with bilateral agreements between the United States and the United Kingdom that were later copied among other nations." (Foust, 2020) The Accord was originally signed on 13th October 2020 with the space agencies of eight States (Prema et al., 2024). The Accord is open for signature indefinitely, with more States expected to join. Until recently, forty-three countries have signed the accord (Artemis Accords, 2024). The Accord reflects "a shared vision of the principles based on the 1967 Outer Space Treaty to create a safe and transparent environment conducive to exploration, science, and business for all mankind to enjoy." (Bardan, 2023).

Although the Accord aims to create a conducive regime for exploration and benefit all of humankind, it is widely conceived as the U.S. policy to utilise and extract space resources unilaterally (Bhukya, 2022). This is unsurprising considering the liberalised licensing policy in the U.S., which is reflected in the 2015 United States Commercial Space Launch Competitiveness Act, recognising private rights over space resources. Also, this was the first instance wherein a domestic law legitimised private entities to claim property rights over space resources (DePagter, 2023). Additionally, President Trump's Executive Order (EO) highlights the U.S. policy towards the commercialisation of space resources (Office of Space Commerce). The EO, in clear terms, reiterates that the U.S. doesn't view space as a global common (Office of Space Commerce). Other States adopted the approach taken by the U.S.; for instance, Luxembourg adopted the 'Law on Use of Resources in Space Act', which permits the appropriation of space resources (Ünüvar, 2022).

States like Japan and the UAE followed suit by adopting similar domestic laws (Foust, 2021; Xu et al., 2023). As Morgan M. DePagter alludes to, these domestic laws operate in distinct ways (DePagter, 2023). This is illustrated by the fact that "Luxembourg, Japan, and the U.A.E. all include the requirement of a license or a permit to retrieve space resources within their statutes." (DePagter, 2023). China and Russia officially announced their plans to set up a lunar base on the Moon called the International Lunar Research Station ('ILRS'); other States can join the project through negotiation with China and Russia (Wall, 2022). Moreover, China and

Russia have criticised the Accord for being tilted in favour of the U.S. Russia compares Trump's space mining order to colonialism (Daemmrich, 2020). These differences could lead to contradictory interpretations regarding the extraction of space resources and their method of law-making, thus abandoning multilateralism. As rightly noted by Rachel Neef, the Artemis Accord and ILRS could create bipolarity in space law in terms of its interpretation (Neef, 2021).

Despite being branded as a U.S. brainchild, the Accord finds favour in developing African and Latin American States (Onwudiwe & Newton, 2021; Esquivel and Manuel, 2024). According to Nigeria's Aerospace Security, the Accord could necessarily fulfil the "socio-economic development of the Nigerian people through investment in science, technology, engineering, and mathematics (STEM) education and earth observation technology (Way, 2020)." These developing States view the Accord as an opportunity to meet their developmental needs.

One of the foremost challenges that the States embracing the Accord needs to address is its compatibility obligation with the Agreement governing the activities of States on the moon and other celestial bodies. Adopted by the United Nations General Assembly on 5 December 1979 (hereinafter referred to as the 'Moon Agreement'). The Moon Agreement is the fifth space treaty developed under the auspice of the United Nations Committee on the Peaceful Uses of Outerspace (COPUOS) and ratified by hardly a handful of States. Despite not getting universal acceptance, the principles embodied in the Moon Agreement, as scholar Laura Delgado-Lopez opines, could resurface because of the emergence of private entities in space activities (Delgado-López, 2015). Companies like Deep Space Industries, Kepler Energy and Moon Express have already made their case for space mining with long-term exploration goals (David, 2023). However, the Accord conflicts with the Moon Agreement by virtue of the former encouraging a pro-commercialization regime, while the latter prohibits commercialisation for individual benefit. This raises questions like whether States are deviating from the object and purpose of the Moon Agreement or whether the Accord constitutes subsequent practice vis-à-vis the Moon Agreement.

In this chapter, the author evaluates these intrinsic tensions through the lens of the Vienna Convention on the Law of Treaties (VCLT), 1969. Accordingly, the author has split the chapter into four parts: the first portion sketches the major principles under the Artemis Accord, and the second segment delves into the divergences between the Accord and the Moon Agreement. Taking forward the discussion, the third part deploys the VCLT to identify and conceptualise the legal nature of the conflicts, and the final part provides the challenges and the conclusion.

5.2 Artemis Accord and the Emerging Principles

In the previous two decades, enormous developments have happened in the space regime, representing a shift in the attitude of the space-faring States (Moltz, 2019). The shift is led by companies motivated by market forces to tap into the rich resources in space. On the Moon, there are copious amounts of Helium-3, a fuel for generating energy (Dobransky, 2013). India's Chandrayaan-1 led the way in the discovery of water molecules on the Moon (Bhandari & Srivastava, 2014). The discovery of water could be a potential source of oxygen, drinking water or water used as rocket fuel for future missions (Kuthunur, 2023). To achieve these objectives, the Accord was drafted under the auspice of the U.S. As Alfonso Augusto Manuel Cocca y Esquivel opines, "[t]he Artemis Accords were created as guidelines for space activities on the Moon and other celestial bodies." (Esquivel and Manuel, 2024) The principles embodied in the Accord take forward the matrix of the Outer Space Treaty (OST), 1967. The OST reflects the Grotian Moment in space law regime; its provisions were drafted in a broad manner to regulate every kind of future contingency. According to Isabel Yolanda Kostadinov, "Instead of rigid phrases and narrow wording, the Treaty is built on overarching principles and contains vague provisions. These legal gaps facilitated the flexible adaption of the Treaty to new developments and evolving technologies (Isabel, 2023)." One of the gaps is reflected in Article II of the OST, which reads: "Outer space, including the Moon and other celestial bodies, is not subject to national appropriation by claim of sovereignty, by means of use or occupation, or by any other means." Private players have pushed for a strict interpretation of Art II, which would imply the non-appropriation principle is aimed solely at the States and not private entities (Gorove, 1969), based on this line of argument, private actors have started selling parts of the Moon.

The principles underlying the Accord can be conceived as a logical extension of the OST and the other four major space law treaties, excluding the Moon Agreement. Despite being considered a political agreement and soft law mechanism, the Accord reflects subsequent development in the wider trend of lawmaking in outer space (Deplano, 2021). Section 3 of the Accord advances that the activities under the Accord should be exclusively for peaceful purposes. This is also drawn from Article IV of the OST. Section 6 on Emergency Assistance seeks to assist personnel in Outer Space who are in distress in line with the obligation under the Rescue Agreement and obligation enshrined under Article V of the OST.

Although several provisions of the Accord are in compliance with pre-existing OST. Section 10 of the Accord was perhaps the most controversial, which pertains to the utilisation and exploitation of space resources (Rivaldi, 2022). Because of the profitability in the extraction of space resources and the gap between the space sector in the U.S. and other states. The U.S. has led a campaign for unilaterally reinterpreting the language of the OST and endorsing the Accord. Article 10 (2) of the Accord reads: "The Signatories emphasize that the extraction and utilization of space resources, including any recovery from the surface or subsurface of the Moon, Mars, comets, or asteroids, should be executed in a manner that complies with the Outer Space Treaty and in support of safe and sustainable space activities. The Signatories affirm that the extraction of space resources does not inherently constitute national appropriation under Article II of the Outer Space Treaty and that contracts and other legal instruments relating to space resources should be consistent with that Treaty."

The Accord favours the extraction of space resources, which brings to question whether space resources are subject to appropriation in the backdrop of the OST and the Moon Agreement. For the purpose of this study, the author has delimited it to the Moon Agreement. In the next section, the author highlights the contradiction between the Accord and the Moon Agreement. This is a bone of contention, especially for States that are simultaneously signatories to both instruments.

5.3 Moon Agreement and the Divergence

Despite a few commonalities between the Accord and the OST, the same does not apply to the Accord and the Moon Agreement. This could have been the reason for the omission of the Moon Agreement from the preamble of the Accord. This intention was made apparent through EO 13914, signed by Donald Trump, in the words of Dr. Scott Pace, this "reaffirms U.S. support for the 1967 Outer Space Treaty while continuing to reject the 1979 Moon Agreement (International Cooperation)."

Unlike the OST, the Moon Agreement is specific in laying out the obligation of the States, leaving very limited room for the State to exploit the gaps in the legal provisions under the OST and, according to Stefan-Michael Wedenig & Jack Wright Nelson, the Moon Agreement elaborates the provision of the OST (Wedenig & Nelson, 2023). For instance, the non-appropriation principle highlighted under Article II of the OST states: "[M]oon is not subject to national appropriation by any claim of sovereignty, by means of use or occupation, or by any other means." The Moon Agreement expands on this provision to include intergovernmental or non-governmental organisations, national entities or any natural persons who are subject to the establishment of an international regime. What follows from the interpretation of Article 11(2) and (3), according to S. Bhat, is that it "forbids private ownership of the moon and other celestial bodies as well as their resources (Bhat, 2007)."

Moreover, unlike the OST, the Moon Agreement has not garnered wide support from States. Currently, only seventeen States are parties to the Agreement. The low figure is because of the non-participation of States like the U.S. and the Soviet Union's vision to explore the Moon in the 1980s (Marboe, 2021). Also, the financial constraint faced by space agencies forced some States to invest in activities capable of getting immediate financial gains, like the development of telecommunication satellites (Tronchetti, 2010). The most striking reason for States to abstain from the Moon Agreement is Article 11, which indicates that the Moon and its natural resources are a Common Heritage of Mankind (CHM) (Rosenfield, 1981). As Fabio Tronchetti opines, "Developed and developing States held different interpretations of this concept and its legal consequences. Due to the hopelessness of reaching a common position on this issue, the space-faring States and the majority of developing countries decided not to become Parties to the Moon Agreement (Tronchetti, 2010)."

When the Argentinian delegate in the Legal Subcommittee presented the draft, it stated that "the natural resources of the Moon and other celestial bodies shall be the

Common Heritage of Mankind (Esquivel and Manuel, 2024)." The principle of CHM had already been mentioned in the drafting of the Law of the Seas Convention (Arikoglu, 2022). The main objection of the U.S. to the principle arose because, in the 1970s, several former colonies became newly independent States, and the idea of a New International Economic Order (NIEO) came into prominence, which aimed at addressing the inequality between the developed and developing States through policies like technology transfer and trade preferences (Mau, 1984). These developments influence the incorporation of CHM in the background of UNCLOS. As Alfonso Augusto Manuel Cocca y Esquivel opines, "[T]he way the CHM was being formulated in the UNCLOS drafts made developed countries apprehensive about it, and this ultimately affected the MOON as well (Esquivel and Manuel, 2024)." The Moon Agreement hitherto remained in cold storage because of a lack of technological advancement in the exploration of the Moon and other celestial bodies. Normative tensions between the Agreement and Accord have resurfaced because of the emergence of private enterprises in the Space regime, which is further backed by the U.S.-led Accord; this raises several normative uncertainties for States signatories to both instruments like Australia.

5.4 Moon Agreement and Artemis Accord: The Normative Divide

Not every aspect of the Accord contradicts the Moon Agreement; there are a few convergences, too. One example pertains to the release of scientific data and assistance to astronauts. In spite of these convergences, there are several blurring points to consider, which may force States to reconsider being part of both instruments. In a broader sense, the Moon Agreement was drafted under the auspice of the United Nations (UN), promoting the idea of multilateralism. Meanwhile, the Accord is a product of unilateralism, reflecting in the words of Tronchetti and Liu the national interest of the U.S. (Tronchetti & Liu, 2021), which is also apparent from the white house EO that buttresses the commercial involvement of private entities (Christensen & Christopher, 2020). Whereas the Agreement is premised on the CHM, the basis of CHM is to share benefits derived from the exploitation of natural resources. As Enver Arikoglu opines, "[T]he attitude of the USA, one of the important space-capable states that openly support private companies exploration of space and the exploitation of its resources through private companies space tourism and Artemis Program, marks the entrance to a new space race era (Arikoglu, 2022)."

On the specific provisions of the Moon Agreement and Accord, there is an opposing view on the utilisation of space resources. The Moon Agreement allows for the commercial use of resources regulated through an international regime. The Moon Agreement requires the State to negotiate for the establishment of an international regime. As David S. Myres argues, the provision reflects institutional regulation pertaining to the Moon and other natural resources (Myers, 1981).

Therefore, signatories to the Moon Agreement joining the U.S.-led initiative could be acting in contravention of this provision. International law requires States to comply with treaty provisions in good faith; this is also echoed in Article 15(2) of the Moon Agreement, which mandates consultation with the State not fulfilling the obligation incumbent upon it and seeking a mutually acceptable resolution.

Ironically, the U.S. appears to have taken a different interpretation; for instance, the U.S. representative Hosenball stated in the backdrop of the Moon Agreement that: "The draft agreement...places no moratorium upon the exploitation of the natural resources on celestial bodies, pending the establishment of an international regime. This permits orderly attempts to establish that such exploitation is, in fact, feasible and practicable by making possible experimental beginnings and, then, pilot operations, a process by which we believe we can learn if it will be practicable and feasible to exploit the mineral resources of such celestial bodies." The breadth of difference related to the peaceful use of outer space is also apparent from the above statement. The Moon Agreement demarcates the limits imposed on military activities. For instance, Article 3(2) of the Agreement reads: "Any threat or use of force or any other hostile act or threat of hostile act on the moon is prohibited. It is likewise prohibited to use the moon in order to commit any such act or to engage in any such threat in relation to the earth, the moon, spacecraft, the personnel of spacecraft or man-made space objects."

This level of specificity is absent in the Accord. The Accord merely states that the activities should be taken for peaceful purposes. Lacking precision in the phrase 'peaceful purposes' could also imply a unilateral interpretation aligning with the U.S. space policy. Fabio Tronchetti views these developments as protecting, defending and projecting American interests, which is concerning (Tronchetti & Liu, 2021). The Accord also provides for the establishment of safety zones, which are essentially designated areas around the spacecraft where activities must be carried out under specific safety regulations (Lea, 2024). The Moon Agreement only makes an indirect reference to the safety zone; Article 9 establishes a lunar station on the Moon, which mandates that the 'station shall use only that area which is required for the needs of the station' and immediately inform the same to the Secretary-General of the United Nations. The bone of contention is that the States parties to the Moon Agreement will not be obligated to recognise the 'safety zone' under the Accord. Also, the Accord provides special consideration to the mitigation of orbital debris, which includes 'the safe, timely, and efficient passivation and disposal of spacecraft at the end of their missions, when appropriate, as part of their mission planning process.' However, there is an absence of any provision pertaining to harmful impact on the Lunar surface, which leads to inconsistency in the Accord and the Moon Agreement.

Therefore, the Accord, although futuristic in its scope by legitimising the utilization of space resources, is at loggerheads with the Moon Agreement on multiple fronts. This could potentially lead to a prospective conflict between the States, especially with the limited resources available on Earth and the States' quest to find an alternative to the Moon. In the next section, the author, through the VCLT, highlights the legal implications of these conflicts.

5.5 VCLT and Special Regimes

There cannot be any legal instrument drafted by man that can be perfect. Hence, interpretation becomes inevitable. The purpose of interpretation is to ascertain the true meaning of the text (Ammann, 2020). According to Georg Nolte, "every application of a treaty presupposes its interpretation" (Notle, 2014). Further, it is also a useful tool to determine the treaty's character, content, effect and scope (Dörr & Schmalenbach, 2018). The VCLT codified the practices of treaty interpretation under Articles 31 and 32, and these provisions have attained the status of customary international law (International Court of Justice, 2004). In the *Kasikili/Sedudu Island case*, although both Namibia and Botswana were not parties to the VCLT, the ICJ observed "… that both of them consider that Article 31 of the Vienna Convention is applicable inasmuch as it reflects customary international law. The Court itself has already had occasion in the past to hold that customary international law found expression in Article 31 of the Vienna Convention (International Court of Justice, 1999)." Also, customary international law rules of treaty interpretation apply to specialised regimes like space law (ILC Report). In the context of WTO, a specialised regime, as aptly put by Mariana Clara de Andrade, recourse to customary international law by an adjudicator is a valuable instrument for understanding the source of international law. Interestingly, Mariana also states that the extent of this reference in WTO is confined to treaty interpretation and state responsibility (de Andrade, 2022). In *Korea-Measures Affecting Government Procurement* (World Trade Organization, 2000), the appellate body discussed the relationship between WTO-covered agreements and general international law. In the words of WTO, "We take note that Article 3 (2) of the DSU requires that we seek within the context of a particular dispute to clarify the existing provisions of the WTO agreements in accordance with customary international law rules of interpretation of public international law (World Trade Organisation, 2000)." Further, Articles 31 and 32 are recognised as customary international law by the WTO. Similarly, specialised regimes like space law are bound by the rule of customary international law and do not operate in total isolation of general international law. General international law can fill the gap in interpreting the ambiguities within the regime. For instance, in Outer Space law, the general principles of state responsibility continue to govern activities in Space (Bhat, 2024). Moreover, States could also take recourse to intuitions like ICJ for settling disputes; the OST lacks a dispute settlement mechanism, and general international law could fill this void. Additionally, the principle of customary international law, including norms under VCLT, operates to facilitate ambiguities in interpreting the OST.

5.6 Artemis Accord and Moon Agreement: Voices from VCLT

Conflict arises between several obligations under the Moon Agreement and the Accord; this is acknowledged by several scholars in the field of space law, but none have addressed the legal implications of the same (Tronchetti & Liu, 2021; de

Zwart, 2021). The VCLT, which applies to special regimes, provides the necessary toolbox to address these conflicts. It is necessary to comprehend the nature of the Accord, as per Article 2(1) (a) of VCLT, "treaty means an international agreement concluded between States in written form and governed by international law, whether embodied in a single instrument or in two or more related instruments and whatever its particular designation." The first requirement is an 'international agreement'. In Draft Article 1, SR Waldock interchangeably uses the phrase 'international agreement' and 'treaties.' (Dörr & Schmalenbach, 2018). In a formal sense, the Accord is an international agreement between sovereign States. Secondly, treaties are concluded between States. States for the purpose of VCLT are flexibly defined, as evident from Article 81 VCLT (popularly defined as the 'Vienna formula). By this rationale, the Holy See became a member of the UN specialised agency, although strictly not fulfilling the criteria of Statehood (Crawford, 2010). Third, the agreement is to be governed by international law, i.e., the governing law is international law. This is to distinguish treaties regulated by international law and those between States regulated by national laws. A state's authority to choose the applicable law is limited to international law, as reflected under Article 38 of the ICJ Statute. Accord is governed by political commitment; this is evident from its preamble. Apart from these requirements under Article 2(1)(a), the States should have the 'intention to establish a legal relationship', which must be capable of establishing the rights and obligations between States, and this distinguishes a political instrument from a legal (Widdows, 2017). The ICJ has identified a few objective criteria for establishing the 'intention'; these are the language of the treaty, negotiating history, and registration under Article 102 of the UN Charter (International Court of Justice, 1962).

The Accord is not registered formally with the UN; instead, as per Article 13 of the Accord, the U.S. government maintain the original text of the Accord and transmits the same to the Secretary-General of the UN. Also, the language of the Accord is couched on specific principles and guidelines rather than creating a set of legally binding obligations. The terms 'intent', 'note', 'commit' and 'recognise' in the Accord reflect the same. The competence to sign any treaty is to be exercised through 'full powers', who could exercise 'full powers' is left to the internal laws of States. The 'full power' emanates from the head of the State, the head of the government and the foreign minister (Bąkowski, 2021). In the case of the Accord, it was signed by members of space agencies, ministers under various departments and the Chief Executive of the 71st International Astronautical Congress. Therefore, incompetency could also be on the basis of the person not being authorised or because a person is acting beyond his mandate.

The Accord, as per the VCLT definition, is not a treaty within the scope of international law and is a mere expression of political commitment. States, through the Accord, may justify their foreign policy objective in commercialising the Moon. However, there could arise two possible scenarios for States signatories to both the Accord and the Moon Agreement; firstly, the Accord may constitute subsequent practice under Article 32 VCLT. Assuming subsequent practice is capable of modifying the treaty, a likelihood of modification can be made through interpretation without an agreement. Interpretation could lead to modification, which might contradict or extend the treaty rule (Arato, 2013). This would endanger ordinary meaning, in context and in light of the object and purpose, under Article 31 VCLT and affecting treaty stability (Bianchet, 2017). Subsequent practice could be a useful tool in the space law regime, considering that space activities would necessitate the adoption of more specific rules. The Accord certainly cannot be classified as a subsequent practice under Article 31(3)(b), as they are not in the application of a treaty that establishes an agreement between the Parties to the Moon Agreement to interpret its provisions. The agreement under Article 31(3)(b) necessitates the agreement between all parties, for instance, the ICJ in the *Whaling in Antarctica case*, took a narrow approach pertaining to the IWC resolution and did not accord interpretative weightage as it was without the support of all the parties to the 1946 International Convention for the Regulation of Whaling (ICRW).

Moreover, nothing in the provision of the Accord refers to the application of the Moon Agreement. According to Irina Buga, the application is the determination of the consequence following the determination of the meaning of the treaty (Buga, 2018). An instance of this can be seen in the case *of Georgia v Russia*, where Georgia submitted ample documents to prove the existence of a dispute, however, the ICJ rejected these materials as it did not make any reference to racial discrimination under the International Convention on the Elimination of All Forms of Racial Discrimination (CERD), this position is also adopted in the WTO cases.[1] Also, the parties to the Accord did not intend to amend the Moon Agreement through the specific conduct. Hence, for the present discussion, Article 31 (3) (b) is tangential.

The Accord could be classified as a supplementary means of interpretation under Article 32 (Deplano, 2021). Unlike Article 31(3)(b), it need not directly relate to the application of the treaty but can still affect the interpretation of the treaty (Buga, 2018). In the identification of subsequent practice under Article 32, it is important to determine the conduct by one or more of the parties in the application of the treaty, but it could be less than all parties to the treaty after the conclusion of the treaty, the Inter-American Court of Human Rights (IACtHR) in *Hilaire, Constantine and Benjamin et al. v. Trinidad and Tobago* observed that the mandatory death penalty for every act resulting in the death of another person contradicts Article 4(2) of the American Convention on Human Rights. To support this proposition, the court took into account the practice of States in the American countries.

According to Aust, subsequent practice can manifest in different forms (Aust, 2013). It could include judgment, a statement pertaining to the interpretation and application of the treaty; subsequent practice under Article 32 further serves as a means of interpretation of the treaty. As Deplano aptly puts it, "Subsequent practice under Article 32 of the VCLT does not require the agreement of all the parties to a treaty (Deplano, 2021)." The purpose of subsequent practice is to play a supporting

[1] In *US–Zeroing* case, the WTO adopted a strict view, accordingly, "In our view, this is simply not a sufficient basis to draw a conclusion as to whether or not the practice of a particular Member to apply Article 2.4.2 to the assessment of the amount of anti-dumping duties within the meaning of Article 9.

role in treaty interpretation. The list provided under Article 32 is not exhaustive and depends on the eye of the interpreter and by which it could include any relevant material. The only limitation to this view is that it is used only to determine the meaning of the treaty provision, they do not replace the original treaty in question. The Accord has strayed from some of the key provisions of the Moon Agreement. As such, it could be stated that although a few provisions of the Accord overlap with the Moon Agreement, they cannot constitute subsequent practice because of their inability to aid in the interpretation of the Moon Agreement. Moreover, the Accord does not fall within the definition of a Treaty under VCLT. Therefore, conflict-avoidance techniques are also inapplicable. Additionally, the Accord reflects the contemporary practice of a few States wanting to assert their hegemony in Outer Space. However, an argument is also being made that the Accord may contribute to the emergence of customary international law through State practice (Smith, 2021). To identify customary international law, two requirements are to be fulfilled, i.e., state practice and *opinio juris* (Baker, 2016).

As the ILC in Draft Conclusion 4 states: "…customary international law refers primarily to the practice of the State…" Opinio Juris, which is subjective or the psychological element, is the belief that the practice is accepted by law (Alcala, 2021). The practice of the State requires to be done with the sense of conviction, as noted by the ICJ in the *North Sea Continental Shelf Case*: "belief that this practice is rendered obligatory by the existence of a rule of law requiring it (International Court of Justice, 1969)." Also, as highlighted in the *Fisheries Case,* "they may be easily understood in the light of the variety of facts and conditions prevailing in the long period (International Court of Justice, 1951)." Additionally, the practice is required to be widespread, representative and consistent. Currently, forty-three States are signatories to the Accord, which does not indicate widespread State practice; further, States like China and Russia have expressed their reservations, too. Russia views the Accord as a "blatant attempt to create international space law that favours the United States (Boley & Michael, 2020)." China travelled the extra mile in equating the move to European colonial enclosure land-taking methods (Elliot et al., 2020). There is a likelihood of subsequent practice crystalising into customary international law and thereby displacing specific treaty provisions.

The VCLT provides limited guidance on the conflict between the treaty and customary international law; in such a scenario, scholars have argued for harmonious interpretation.[2] Specific rules of treaty interpretation in such cases have to be evaluated from the perspective of the subject matter and the will of the parties to establish *lex specialis*, *lex posterior*, *lex prior* and *lex superior* by means of interpretation (Buga, 2018). However, there is a presumption against the prospect of a treaty modifying a customary law and vice-versa. In future, assuming the Accord provisions like disclosure of scientific data, safety, transparency, and safety in Outer Space attain the status of customary international law, it could possibly modify specific provisions of the Moon Agreement; such instances of external sources falling under

[2] The rule of systemic integration, as reflected in Article 31(3)(C) VCLT.

Article 32 constitute external customary international law. In such instances, as Irina Buga points out, the evidentiary weightage of the practice is lower than subsequent practice under Article 31(3)(b); since the Accord squarely falls under Article 32, its evidentiary weightage is limited (Buga, 2018). On top of that, the Accord is a recent unilateral move by the U.S. and has only started to gain ground.

Secondly, the States could violate the object and purpose of the Moon Agreement by virtue of being a signatory to the Accord. Article 18 of VCLT lays down the State's obligation not to defeat the object and purpose of the treaty prior to its entry into force. Despite States signing the Moon Agreement, many States have yet ratified it, therefore, the Moon Agreement has not entered into force vis-à-vis the said States. The interim obligation does not necessitate full compliance with the treaty, but the object and purpose are to be respected. This interim obligation is to protect the legitimate expectation of the other States participating in the treaty-making process. The judicial bodies have attempted to define the object and purpose. In the Genocide Reservations Case, the ICJ attempted to maintain a balance between treaty integrity and permitting reservation as a sovereign entitlement. However, as the ICJ made it evident that, "if a party to the Convention objects to a reservation which it considers to be incompatible with the object and purpose of the Convention, it can in fact consider that the reserving State is not a party to the Convention." (International Court of Justice, May 1951) Interestingly, the ICJ did not clarify what constitutes the object and purpose of a treaty. The International Law Commission (ILC) work on Reservation provides guidance in terms of the determination of the object and purpose, according to guideline 3.1.5.1,

"The object and purpose of the treaty is to be determined in good faith, taking account of the terms of the treaty in their context, in particular the title and the preamble of the treaty. Recourse may also be had to the preparatory work of the treaty and the circumstances of its conclusion and, where appropriate, the subsequent practice of the parties." The ILC work has identified four criteria to identify the object and purpose: first, the determination to be done in good faith; second, recourse to the terms of the treaty, particularly in the title and preamble; third, the preparatory work; and fourth, subsequent practice. As the ILC itself acknowledges, it is not easy for these elements to operate together to determine the object and purpose, in the words of the ILC, "Such a process undoubtedly requires more esprit de finesse than esprit de géométrie, like any act of interpretation, for that matter—and this process is certainly one of interpretation (Kritsiotis, n.d.)." Klabbers urges the use of 'common sense' in identifying the object and purpose (Klabbers, 1997).

The object and purpose of the Moon Agreement is to find a balance between 'exploration and exploitation'. (Khubchandani, 2023) Individual commercialisation of the Moon, although would be in line with the provision of the Accord, would contradict the provisions of the Moon Agreement. Article 4(1) of the Moon Agreement notes: "The exploration and use of the moon shall be the province of all mankind and shall be carried out for the benefit and in the interests of all countries, irrespective of their degree of economic or scientific development." The provision is linked to Article 11 of the Moon Agreement, which, in the words of Antonella Bini, is the most important provision of the Agreement (ESPI). Both the provisions reflect the object and purpose of the Moon Agreement, this is inferred from the preamble of the Moon Agreement, "...to promote on the basis of equality the further development of co-operation among States in the exploration and use of the moon and other celestial bodies." Further, as highlighted in the preamble, the Moon Agreement is intended to develop and define the provisions of the five core treaties on Outer Space law. Accordingly, Article 11 (5) was intended to develop an international regime to regulate the exploitation of the Moon; this is as per the doctrine of CHM, a concept first developed in the backdrop of the law of the seas in Part XI of UNCLOS to distribute benefits in the deep seabed. Hence, the extraction and utilisation of space resources and claiming them as not constituting national appropriation hit directly at the object and purpose of the Moon Agreement.

Additionally, Article 11 of the Accord provides for the establishment of a 'safety zone' to prevent harmful interference. This safety zone has to be temporary and provide free access to all areas of the Moon. Only time will tell if these safety zones are zones for indirect appropriation. Article 9 of the Moon Agreement talks about the establishment of Moon stations; the area shall be only used for the needs of the station. Therefore, there is a possibility that the States party to the Moon agreement will not recognise the legality of the safety zone. In these circumstances, States are bound to prioritise the legal obligation under the Moon Agreement over their political commitment; this stems from the 'good faith' obligation. This is further backed by Article 31 (1) of VCLT. Scholars have considered good faith as one of the most important principles in modern times, this is also reflected in the *Nuclear Test Case*, where the ICJ observed: "One of the basic principles governing the creation and performance of legal obligations, whatever their source, is the principle of good faith (International Court of Justice, 1974)."

The law-making exercise in Outer Space law is moving in the direction of soft law mechanism over the traditional treaty method, perhaps the Accord signals this detour. Although soft laws don't create law, they contribute to the consolidation of international law (Pronto, 2021). Efforts have been taken to formulate multilateral treaties in 2008 and 2014 through the submissions of China and Russia pertaining to the Draft Treaty on the Prevention of the Placement of Weapons in Our Space and of the Threat of Force against Outer Space Objects (FMPRC, 2008). These proposed treaties were of limited success in establishing binding obligations. The Accord stands mid-way in terms of multilateralism and national legislation and could play a prominent role in interpreting the law. There could be two possible ways out; States could opt out of the Moon Agreement or the Accord, like Saudi Arabia, which backed out of the Moon Agreement; the reasons are seen as economic and strategic. Second, specific provisions of the Moon Agreement could be amended to ensure its compliance with the Accord. This can be proposed by any State party to the Agreement, which shall enter into force for each State accepting the amendments. Alternatively, an implementing agreement could supplement the Moon Agreement, enhancing the specific procedure relating to exploitation and harmonizing the principles of the Moon Agreement and the Accord.

5.7 Conclusion

The Moon Agreement certainly needs a relook in the backdrop of the emerging private actors in the space sector. However, this process is to be carried out through a multilateral framework, where the aspirations of different States are taken on board via international cooperation. The Accord challenges the fundamental tenets of multilateralism and international cooperation. Further, the Accord's compatibility with the Moon Agreement is also contested. Interpretative techniques in the VCLT could expose some of these faultlines. This research establishes the application of VCLT in interpreting the Accord vis-à-vis the Moon Agreement is finite. States concomitantly parties to the Moon Agreement and Accord could violate the object and purpose of the former. From an interpretative lens, States are bound to respect the object and purpose of the treaty even prior to the treaty entering into force in good faith. Besides, the Accord does not constitute subsequent practice under VCLT; hence, it's not an instrument to interpret the Moon Agreement. Instead, some key concepts in the Accord may crystalise into customary international law, but that is left to prospective developments in space law jurisprudence.

References

Alcala, R. (2021). *Opinio juris and the essential role of states.* Available at <https://lieber.westpoint.edu/opinio-juris-essential-role-states/>

Ammann, O. (2020). The interpretative methods of international law: What are they, and why use them? In *Domestic courts and the interpretation of international law* (pp. 191–222). Brill Nijhoff.

de Andrade, M. C. (2022). Identification of and resort to customary international law by the WTO appellate body. In P. Merkouris, J. Kammerhofer, & N. Arajärvi (Eds.), *The theory, practice, and interpretation of customary international law. The rules of interpretation of customary international law* (pp. 277–297). Cambridge University Press.

Arato, J. (2013). Treaty interpretation and constitutional transformation: Informal change in international organizations. *Yale Journal of International Law, 38*, 289.

Arikoglu, E. (2022). The concept of common heritage of mankind and legal status of outer space in international law. In B. E. Balin, V. N. Akun, & S. Alis (Eds.), *Proceedings for the first symposium on space economy, space law and space sciences* (pp. 109–116). Istanbul University.

Artemis Accords, Principles for Cooperation in the Civil Exploration and Use of the Moon, Mars, Comets, and Asteroids for Peaceful Purpose'. (2024) available at <https://www.state.gov/artemis-accords/>

Aust, A. (2013). *Modern treaty law and practice* (3rd ed.). Cambridge University Press.

Baker, R. R. B. (2016). Customary international law: A reconceptualization. *Brooklyn Journal of International Law*, 41. Available at <>https://brooklynworks.brooklaw.edu/bjil/vol41/iss2/1>

Bąkowski, M. (2021). The institution of full powers in the process of concluding international agreements. *Przegląd Prawniczy Uniwersytetu im. Adama Mickiewicza, 13*, 97–117.

Bardan, R. (2023). *NASA Sets Coverage for Czech Republic Artemis Accords Signing Ceremony*. Available at <https://www.nasa.gov/news-release/nasa-sets-coverage-for-czech-republic-artemis-accords-signing-ceremony/>

Bhandari, N., & Srivastava, N. (2014). Active moon: Evidences from Chandrayaan-1 and the proposed Indian missions. *Geoscience Letters, 1*, 11.

Bhat, S. (2007). The concept of common heritage of mankind in the governance of the moon – An insight into article 11 of the moon agreement. *Legal Opus.*

Bhat, B. S. (2024). State responsibility and international liability. In B. B. Sandeepa, D. Ukey, & A. Variath (Eds.), *International space law in the new space era: Principles and challenges.* Oxford Academic.

Bhukya, K. (2022). *The artemis accords: A turn towards international cooperation or a weapon for establishing U.S. hegemony?.* Available at <https://www.pulj.org/the-roundtable/the-artemis-accords-a-turn-towards-international-cooperation-or-a-weapon-for-establishing-us-hegemony>

Bianchet, L. (2017). *Treaty Modification by Subsequent Practice: Analysis of the present legal framework, the expansive potential and the consequences of treaty modification by subsequent practice.* Available at <https://www.diva-portal.org/smash/get/diva2:1189907/FULLTEXT02>

Boley, A., & Michael, B. (2020). *U.S. policy puts the safe development of space at risk.* Available at <https://ui.adsabs.harvard.edu/abs/2020Sci...370..174B/abstract>

Buga, I. (2018). Subsequent customary law as a means of treaty modification. In *Modification of treaties by subsequent practice.* Oxford Academic.

Christensen, I. A., & Christopher, D. J. (2020). *Putting the White House executive order on space resources in an international context.* Available at <https://www.thespacereview.com/archive/3932-1.html>

Cocca y Esquivel, Alfonso Augusto Manuel. (2024). 'The Moon Agreement and the Artemis Accords: Is Latin America at a Crossroads?' *Space Fostering Latin American Societies: Developing the Latin American Continent Through Space, Part 5.* (Cham: Springer Nature Switzerland).

Crawford, J. (2010). The criteria for statehood applied: Some special cases. In *The creation of states in international law.* Oxford Academic.

Daemmrich, B. (2020). Russia compares Trump's space mining order to colonialism. *The Moscow Times.* Available at <https://www.themoscowtimes.com/2020/04/07/russia-compares-trumps-space-mining-order-to-colonialism-a69901>

David, L. (2023). *Space mining startups see a rich future on asteroids and the moon.* Available at <https://www.space.com/space-mining-grinding-into-reality>

Delgado-López, L. (2015). Beyond the Moon Agreement: Norms of responsible behavior for private sector activities on the moon and celestial bodies. *Space Policy, 32*, 6–8.

DePagter, M. M. (2023). Who dares, wins: How property rights in space could be dictated by the countries willing to make the first move. *Chicago Journal of International Law*, CJIL Online 1.2.

Deplano, R. (2021). The Artemis Accords: Evolution or revolution in international space law? *International & Comparative Law Quarterly, 70*(3), 799–819.

Dobransky, S. (2013). *Helium-3: the Future of energy security.* Available at <https://go.gale.com/ps/i.do?id=GALE%7CA326001921&sid=googleScholar&v=2.1&it=r&linkaccess=abs&issn=07423640&p=AONE&sw=w&userGroupName=anon%7E6a25cb1&aty=open-web-entry>

Dörr, O., & Schmalenbach, K. (2018). *Vienna convention on the law of treaties.* Springer.

Elliot, J., Cerny, M. B., & Piliero, R. J. (2020). *What does China think about NASA's artemis accords?.* Available at <https://thediplomat.com/2020/09/what-does-china-think-about-nasas-artemis-accords/>

ET Bureau. (2020). *NASA all set to send the first woman to moon in its next planned manned mission in 2024.* Available at <https://economictimes.indiatimes.com/magazines/panache/nasa-all-set-to-send-the-first-woman-to-moon-in-its-next-planned-manned-mission-in2024/articleshow/79690608.cms?from=mdr>

FMPRC, China and Russia jointly submitted the draft Treaty on PPWT to the Conference on Disarmament. (2008). Available at <https://www.fmprc.gov.cn/mfa_eng/wjb_663304/zzjg_663340/jks_665232/jkxw_665234/200802/t20080212_599177.html>

Foust, J. (2020). *Eight countries sign Artemis Accords.* Available at <https://spacenews.com/eight-countries-sign-artemis-accords/>

Foust, J. (2021). *Japan passes space resources law*. Available at https://spacenews.com/japan-passes-space-resources-law/

Gorove, S. (1969). Interpreting article II of the outer space treaty. *Fordham Law Review, 37*(3), 349.

Grush, L. (2020). *US and seven other countries sign NASA's Artemis Accords to set rules for exploring the Moon*. Available at <https://www.theverge.com/2020/10/13/21507204/nasa-artemis-accords-8-countries-moon-outer-space-treaty>

Isabel, Y. K. (2023). *Colonisation of space including the moon and other celestial bodies: Is there a legal regime?*. Available at https://core.ac.uk/download/553555236.pdf

Khubchandani, M. (2023). *As the First Country to Land on the Moon's South Pole, Should India also be the First Space Power to Ratify the Moon Agreement?* Available at https://www.ejiltalk.org/as-the-first-country-to-land-on-the-moons-south-pole-should-india-also-be-the-first-space-power-to-ratify-the-moon-agreement/

Klabbers, J. (1997). *Some problems regarding the object and propose of treaties*. 8 FinnYIL.

Kritsiotis, D. (n.d.). The object and purpose of a treaty's object and purpose. *Conceptual and Contextual Perspectives on the Modern Law of Treaties*, 237–302.

Kuthunur, S. (2023). *Scientists find hydrogen in Apollo moon rocks, suggesting astronauts can harvest lunar water*. Available at <https://www.space.com/hydrogen-moon-rocks-apollo-astronauts-sample>

Lea, R. (2024). *Artemis Accords: What are they & which countries are involved?* Available at <https://www.space.com/artemis-accords-explained>

Malik, T. (2023). *NASA may build more than one moon base for Artemis lunar missions*. Available at <https://www.space.com/nasa-artemis-base-camp-more-moon-bases>

Marboe, I. (2021). What, if any, relevance does the Moon Agreement have to activities in space today? *The New Space Age: Beyond Global Order*. Available at <https://global.upenn.edu/sites/default/files/perryworldhouse/Marboe_SpaceWorkshop.pdf>

Mau, S. D. (1984). Equity, the third world and the moon treaty. *Suffolk Transnat'l Law Journal, 8*.

Moltz, J. C. (2019). The changing dynamics of twenty-first-century space power. *Journal of Strategic Security, 12*(1), 15–43.

Myers, D. S. (1981). The moon treaty in legal and political perspective. *Proceedings of the Twenty-third Colloquium on the Law of Outer Space*, 49–55.

NASA. (2020). *NASA names companies to develop human landers for artemis moon missions*. Available at <https://www.nasa.gov/news-release/nasa-names-companies-to-develop-human-landers-for-artemis-moon-missions/>

Neef, R. (2021). Artemis accords: A new path forward for space lawmaking? *Adelaide Law Review, 42*, 569.

Nolte, G. (2014). *Second report on subsequent agreements and subsequent practice in relation to the interpretation of treaties* (U.N. Doc. A/CN.4/671). United Nations.

Onwudiwe, M., & Newton, K. (2021). Africa and the artemis accords: A review of space regulations and strategy for african capacity building in the new space economy. *New Space, 9*(1), 38–48.

Prema, E., Bharathi, R. V., & Sundar, V. S. (2024). The artemis accords: A new phase in the development of outer space law. In *Ascending to space: Critical perspectives from New Zealand and other nations* (pp. 73–97). Springer Nature.

Pronto, A. N. (2021). Understanding the hard/soft distinction in international law. *Vanderbilt Law Review, 48*, 941.

Rivaldi, R. (2022). The artemis accords and property rights in outer space. *Journal of Law and Policy Transformation, 7*(2), 36–46.

Rosenfield, S. B. (1981). The moon treaty: The United States should not become a party. In *Proceedings of the ASIL annual meeting* (Vol. 74). Cambridge University Press.

Smith, W. A. (2021). Using the artemis accords to build customary international law: A vision for a U.S.-centric good governance regime in outer space. *Journal of Air Law and Commerce, 86*(4), 661.

Tronchetti, F. (2009). *The exploitation of natural resources of the Moon and other celestial bodies: A proposal for a legal regime*. Martinus Nijhoff Publishers.

Tronchetti, F. (2010). The moon agreement in the 21st century: Addressing its potential role in the era of commercial exploitation of the natural resources of the moon and other celestial bodies. *Journal of Space Law, 36*(2), 489–524.

Tronchetti, F., & Liu, H. (2021). Australia's signing of the Artemis Accords: A positive development or a controversial choice? *Australian Journal of International Affairs, 75*(3), 243–251.

Ünüvar, G. (2022). *Can national laws on space resources serve as evidence of customary international law?*. Available at <http://opiniojuris.org/2022/11/08/can-national-laws-on-space-resources-serve-as-evidence-of-customary-international-law/>

Wall, M. (2022). *Not just Artemis: China and Russia plan to put boots on the moon, too*. Available at <https://www.space.com/china-russia-moon-base-ilrs>

Way, T. (2020). *Challenges and opportunities of Nigeria's space program*. Available at <https://aerospace.csis.org/challenges-and-opportunities-of-nigerias-space-policy/>

Wedenig, S.-M., & Nelson, J. W. (2023). *The moon agreement: Hanging by a thread?* Available at <https://www.mcgill.ca/iasl/article/moon-agreement-hanging-thread>

Widdows, K. (2017). What is an agreement in international law? In *The Law of Treaties* (pp. 73–105). Routledge.

World Trade Organization, Korea – Measures Affecting Government Procurement, Report of the Panel, WT/DS163/R 1 May 2000. Available at https://www.worldtradelaw.net/document.php?id=reports/wtopanelsfull/koreagovprocurement(panel)(full).pdf&mode=download#page=2>

Xu, F., Su, J., & Mehdi, M. (2023). A re-examination of fundamental principles of international space law at the dawn of space mining. *Journal Of Space Law, 44*, 1.

de Zwart, M. (2021). To the moon and beyond: The Artemis Accords and the evolution of space law. *Commercial and military uses of outer space*, 65–80.

Chapter 6
Increasing Fragmentation in Regulating Space Resources: The Imperative of Returning to a Multilateral Approach

Athar Ud Din

Abstract With the advancement of technology, the exploration and use of space resources have captured the attention of various stakeholders, necessitating the need for an unambiguous regulatory framework. However, the existing international law relating to the exploration and use of outer space does not provide sufficient clarity relating to the ownership and use of space resources. This has resulted in increasing fragmentation of space law, especially relating to developing a regulatory framework for space resources. Plurilateral arrangements like the Artemis Accords, accompanied by national space policies for exploration of outer space, further contribute to this fragmentation, thereby undermining the emergence of an UN-guided multilateral approach that prioritises coherence and inclusiveness. Amidst the ongoing debates over possessory rights in outer space, this chapter asserts that while the substantive outcome of a regulatory framework is often prioritised, the process of creating the regulatory framework is equally important. More specifically, the paper emphasises the need to adopt a multilateral approach, specifically under the ambit of the United Nations, to establish a well-defined and equitable regulatory approach for space resources. By examining the recent developments in India's space program and policy and ongoing work at COPUOS, this chapter contends that emerging powers should actively engage in COPUOS to help develop a more collaborative and equitable framework for space resources.

Keywords Space resources · Artemis Accords · Fragmentation · International space law · National space policies

A. Ud Din (✉)
Centre for International Relations, Islamic University of Science and Technology, Awantipora, India
e-mail: athar.uddin@islamicuniversity.edu.in

S. Bhat B., A. Variath (eds.), *Artemis Accords and Resource Mining in Outer Space*, Issues in Space, https://doi.org/10.1007/978-981-95-0152-6_6

6.1 Introduction

The perception of outer space is rapidly evolving, moving beyond its traditional association with the awe inspired by its vastness and mysteries. Increasingly, outer space is being visualised through a utilitarian lens, as a lucrative domain fit for human exploration and exploitation. In this context, it is important to take note of key recent developments, which include the formulation of national space laws and policies, the establishment of the Artemis Accords, and India's Moon landing at the south pole of the Moon. Moreover, future initiatives such as space tourism, building settlements on planets like Mars, and asteroid mining are on the horizon.

In light of these advancements, States need to develop and adapt international law to respond to the rapid technological and political landscape. One report suggests that mining the top ten asteroids (based on their proximity to Earth and value) could yield a profit of US$1.5 trillion (Yarlagadda, 2024). Further, the worth of gold in one asteroid (*Psyche 16*) is enough to give each person on Earth US$93 billion (Yarlagadda, 2024)! However, the vastness of outer space and the associated technological costs make resource extraction in outer space extremely costly. It means that space resources are a scarce commodity (Tjandra, 2021, p. 391). This only accentuates the need to have a clear, well-defined and equitable regulatory framework for resource extraction and possession in outer space. Within the field of international space law, the nature of possessory rights over space resources will be of critical significance because of the existing regulatory gaps.

Amidst the debates relating to the ideal framework for possessory rights in outer space, this chapter contends that the primary focus should be on the process through which decisions are made. Procedural aspects of decision making have gained more importance because of recent developments that include the signing of the Artemis Accords, India's Moon Landing, the adoption of national space laws and policy, the growing proximity between Russia and China for space exploration, and the recent developments at COPUOS. The growing fragmentation of outer space law threatens to extend existing territorial tensions to outer space, making procedural aspects of international rulemaking relating to outer space essential to ensuring that future space exploration is carried out in the spirit of cooperation, instead of confrontation (Tjandra, 2021, p. 391).

The chapter is organised as follows: the first section provides an overview of regulatory gaps for possessory rights in outer space, along with associated developments in national space policies and laws of important space-faring countries. Subsequently, the chapter focuses on the Artemis Accords and their impact on the multilateral governance of outer space. The final part of the chapter focuses on India's Moon landing and recent trends in its space policy. It contends that countries like India, along with other emerging powers, could play a decisive role in shaping the procedural dimensions of international space law.

6.2 Managing Space Resources: Regulatory Gaps and Growing Fragmentation

The existing international space law is primarily founded on the *Treaty on Principles Governing the Activities of States in the Exploration and Use of Outer Space, including the Moon and Other Celestial Bodies*, which entered into force in 1967 (Outer Space Treaty, 1967a). The Outer Space Treaty establishes the foundational principles for space exploration, supplemented by a series of additional treaties that further elaborate upon the basic and foundational principles. Two aspects of the existing international space law are worth taking into account: the dates of its formulation and the design it follows. Regarding the dates, it is important to note that the international space law came into being in the period ranging from 1967 to 1979, with the Moon Agreement of 1979 marking the last major treaty in this area. These dates do not encompass the Artemis Accords, which cannot be equated to international treaties. In any case, the Artemis Accords claim to be grounded in the principles established in the Outer Space Treaty and thus build upon its established framework (NASA, 2024). In other words, the outer space law was formulated in a period when the USA and the USSR were the two major space-faring powers. Furthermore, the development of outer space law was shaped by the technological context and geopolitical considerations of that era. In other words, the international space law developed against the backdrop of the Cold War, with a focus on preventing militarisation of outer space, and keeping in account the technological advancements of the time (Zhao, 2004, p. 283).

The other noteworthy aspect of international space law is its structural design. Similar to some other areas of international law, such as law relating to climate change, international space law is based on a framework in which a foundational agreement establishes core principles, which are further expanded through treaties. The Outer Space Treaty serves as the foundational agreement setting forth the basic principles, which are then elaborated upon in subsequent treaties.

6.2.1 International Law and the Ownership and Use of Space Resources

Space law evolved in an era of a different geopolitical and technological landscape. Accordingly, extraction, ownership and commercial use of space resources do not find explicit reference in the Outer Space Treaty. Article I of the Outer Space Treaty enshrines that all states are free to explore and use outer space, and the exploration and use of outer space shall be the 'province of all mankind'. Article II prohibits claims of national appropriation of outer space, which includes celestial bodies. Unlike the United Nations Convention on the Law of the Sea (UNCLOS), the Outer Space Treaty does not mention resources contained in the celestial bodies. Although no explicit rule relating to ownership of space resources exists in the Outer Space

Treaty, at the very least, states are under an obligation to ensure not to use outer space in a manner that deprives other states of the use and benefits of outer space (Lele & Gopalakrishnan, 2020).

While Outer Space Treaty does not shed light on the ownership of space resources, the Moon Agreement of 1979 lays down details relating to the legal framework that would govern exploration and use of space resources (Moon Agreement, 1979). It follows the pattern of the UNCLOS and provides that labels that the Moon and its resources are a Common Heritage of Mankind (CHM). The exact scope of CHM is contested, but a broad understanding of the principle includes the following: prohibition of national appropriation; peaceful use; joint management through an international regime; and equitable sharing of benefits (Noyes, 2012).

Accordingly, the Moon Agreement envisions the establishment of an international regime for regulating the use of resources of the Moon and ensuring equitable sharing of benefits, with special consideration given to developing countries and those states that have contributed to the exploration of the Moon. Moreover, Article 1 of the Moon Agreement provides that "The provisions of this Agreement relating to the moon shall also apply to other celestial bodies within the solar system, other than the Earth, except in so far as specific legal norms enter into force with respect to any of these celestial bodies." In other words, till a specific legal regime is adopted for other celestial bodies (asteroids, Mars included) the legal arrangement for the resources of the Moon extends to other space resources as well.

Given the design of international space law, could the Moon Agreement be considered an elaboration of Articles I and II of the Outer Space Treaty? This argument could have been plausible had it not been for the fact that the Moon Agreement has not been received favourably. So far, only 17 states have ratified it, and the list does not include major space-faring powers. Therefore, the silence of the Outer Space Treaty, coupled with limited ratification of the Moon Agreement, implies that the international space law does not adequately and explicitly address the question of ownership of space resources. Perhaps, it is not surprising, given the geopolitical and technological context of the era when the law of outer space was being negotiated. The regulatory gap is being filled with scholarly arguments and national space policies.

As far as scholarly arguments are concerned, they vary widely, ranging from calls for outright rejection of any form of communitarian approach to property rights in outer space (Thomas, 2005), to proposals that allow appropriation provided certain reserves are set aside for the use and exploitation of developing countries (Merges & Reynolds, 1997). One possible argument could be that the framework adopted in the law of the law of sea should be used for the management of resources in outer space. This is because drafters of international space law were cognizant of the developments in *United Nations Conventions on the Law of the Sea* (UNCLOS) regarding equitable sharing of benefits from resources outside the territorial jurisdiction of states (Committee on Peaceful Uses of Outer Space, 1979). Another approach is to differentiate between celestial bodies and the resources contained therein, with only the former being subject to prohibition of claims of non-appropriation (Melchin, 2015; Christol, 1980, p. 430). Evidently, proposals for appropriation of space resources on a 'first come, first served' basis- effectively treating celestial bodies as *res nullius* carries significant risks. Further, as stated by the then Malta's representative to the UN, Mr. Arvid Pardo, there are instances of states asserting ownership of distant territories, even without establishing effective control, or on the grounds of technological advancements (United Nations, 1967b, para. 58). In this context, it is important to mention that many countries have adopted national space policies that allow appropriation of outer space resources. An overview of national space policies and laws is provided below.

6.2.2 Growth of National Space Policies

Against the backdrop of regulatory gap relating to ownership of space resources, many countries have adopted national space policies and laws. Of particular relevance to the current discussion is the fact that, with a few exceptions, such as India, these national space policies and laws clearly articulate their position on the extraction and use of space resources. The section below provides a brief overview of selected national space laws and policies.

The US policy has always been unequivocal about the ownership of resources situated outside its territorial jurisdiction. Even in the case of the law of the sea, the USA has interpreted the CHM principle as analogous to a common pool, from which all can draw resources (Goldie, 1983, p. 105). It is not surprising, then, that the *National Space Policy of the United States of America* asserts that international law relating to outer space does not impose restrictions on the activities of states (United States, 2010). This position is further clarified in the *U.S. Commercial Space Launch Competitiveness Act*, which distinguishes between celestial bodies and their natural resources, thereby permitting commercial exploitation of space resources by US citizens (U.S. Commercial Space Launch Competitiveness Act, 2015, § 513.02.a). Furthermore, the USA is the main architect and driver of the Artemis Accords, which, as the subsequent sections will demonstrate, adopt a similar position on ownership of space resources. In 2016, Luxembourg launched the 'SpaceResources.lu Initiative' aimed at exploration and use of space resources (Luxembourg Space Agency, 2019). The UAE has also adopted a regulatory framework for space resources that contains provisions for the commercial use of space resources (UAE Space Agency, 2023, Art. 7).

Notably, China in its submission to the UN Committee on the Peaceful Uses of Outer Space stated that the Outer Space Treaty is the "cornerstone of the existing international space law", and rules relating to space resources should be formulated on the basis of the Outer Space Treaty (United Nations Office for Outer Space Affairs, 2024, para. 4). More significantly, China in its submission stated: "Article II of the Outer Space Treaty provides that outer space, including the Moon and other celestial bodies, is not subject to national appropriation by claim of sovereignty, by means of use or occupation, or by any other means. The Chinese Delegation believes that the exploration, exploitation and utilisation of space resources shall be in conformity with the principle of non-appropriation…" (Jones, 2024).

It is essential to mention that at the COPUOS has published *the Initial draft set of recommended principles for space resource activities* on 27th of March 2025 (COPUOS, 2025). Part 1 of the draft principles prohibits any use of space resource activities that constitutes appropriation of outer space or celestial bodies. However, in the specific case of space resources Part 2 provides that extraction and use of space resources does not inherently constitute appropriation. While these are draft principles and yet to be officially adopted in the form of legal instrument, they signify a timely step towards a multilateral approach for managing space resources.

6.3 The Artemis Accords: Wither Multilateralism?

The Artemis Accords were established in 2020 by NASA (the national space agency of the USA), along with the US Department of State, with the objective of providing a common set of principles to govern exploration and use of outer space. Initially, the Accords were established in partnership with seven other initial signatory nations, and since then, space agencies of several other countries have become part of the Artemis Accords. As of April 8, 2025, Bangladesh became the 54th nation to sign the accords. The list of signatory countries transcends the traditional North-South divide in international legal relations.

Two aspects of the Artemis Accords are of specific relevance here: the position on ownership and use of space resources, and the design of the regulatory framework that the Accords aim to establish. In the context of ownership of space resources, Section 10 of the Artemis Accords is of relevance. It provides "The Signatories emphasise that the extraction and utilisation of space resources, including any recovery…should be executed in a manner that complies with the Outer Space Treaty and in support of safe and sustainable space activities." It further provides that "the Signatories affirm that the extraction of space resources does not inherently constitute national appropriation under Article II of the Outer Space Treaty, and that contracts and other legal instruments relating to space resources should be consistent with that Treaty" (NASA, 2020, Section 10, para. 2).

While the first part of Article 10 affirms compliance with the Outer Space Treaty, it is the second part that takes an unequivocal position on the extraction of space resources. It asserts that the extraction of space resources does not violate Section II of the Outer Space Treaty. As already noted above, the Outer Space Treaty does not explicitly state its position on ownership and use of space resources. In other words, Section 10 of the Artemis Accords adopts a definitive position on Section II of the Outer Space Treaty, a position taken and put into effect without the consensus of all states that have ratified the Outer Space Treaty.

Another important aspect of the Artemis Accords is their structural design. The Artemis Accords establish foundational principles that will guide space exploration, with these principles to be further developed through supplementary agreements, such as MoUs and bilateral agreements between the relevant space agencies. This design mirrors the approach followed in international space law, where the Outer Space Treaty contains basic principles, which are elaborated upon by subsequent agreements. Although, Artemis Accords differ from formal treaties in terms of legal status—The Accords are non-binding arrangements, whereas treaties relating to outer space are legally binding and ratified by states—the growing number of signatories to the Artemis Accords, combined with the definitive position the Accords take towards ownership of space resources (in contrast to ambiguity in outer space law), suggests that the Artemis Accords could potentially emerge as an alternate legal framework for space exploration.

Indeed, developments happening elsewhere suggest that the Artemis Accords are met with a degree of scepticism. For instance, in China, there is a growing perception that the Artemis Accords do not comply with the provisions of the Outer Space Treaty (Ji et al., 2020; Wu, 2023). More noteworthy is the news that China and Russia have started to cooperate to create the International Lunar Research Station (Wall, 2022). Their invitation to other countries to join this initiative is indicative of the fact that they envision it to emerge as a competing framework to the Artemis Accords.

Amidst the ongoing competition and efforts to establish alternative legal frameworks for outer space exploration, UN bodies like the United Nations Office of Outer Space Affairs (UNOOSA) and the UN Committee on the Peaceful Uses of Outer Space (COPUOS), which have historically played an important role in a multilateral development of outer space law, appear to be potentially marginalised. Further, those developing countries whose space program is still in lack a meaningful voice in shaping the emerging space law relating to ownership of space resources. Their perspectives are overshadowed by the growing fragmentation and rivalry in pursuit of space resources. In other words, the future of space exploration is increasingly being shaped by geopolitical tensions, as opposed to multilateral processes within the United Nations. It is essential to mention here that the draft principles being developed at COPUOS do not necessarily conflict with Artemis Accords as far as ownership of space resources is concerned. However, it is essential to prioritize the work done at COPUOS because, of its multilateral and collaborative approach, which, as previously discussed, warrants greater emphasis.

6.4 The Distinctive Position of India

India has long been recognised as a country with significant potential for space exploration. India's national space agency, ISRO, has demonstrated its prowess through important achievements like launching more than 100 satellites from a single rocket (ISRO, 2017) and deployment of orbital missions to distant bodies like Mars (ISRO, 2024a). However, with the recent landing of Chandrayaan-3 on the south pole of the Moon, India has firmly established itself as an important space-faring power (Pandey, 2023). The success of the Chandrayaan-3 mission signifies that India is no longer an emerging space power but has established its position as an important actor in space exploration. More importantly, India framed its Moon landing within the broader context of the achievements of the Global South (Tingley, 2023). Therefore, India's stance and policies towards space exploration carry a significant weightage. In this context, India's space policy and the potential implications of its joining of the Artemis Accords deserve specific attention.

6.4.1 India's Space Policy

For a long time, India did not have a dedicated national space policy and its space policy was governed by the Satellite Communication Policy and Remote Sensing Data Policy. In 2017, India also circulated a *Draft Space Activities Bill* to solicit feedback from various sections of the civil society with the aim of enacting a comprehensive national space law. However, so far, a national space law has not been enacted in India. Instead, India has recently adopted the *Indian Space Policy—2023* (ISRO, 2024b). It aims, among other objectives, to streamline and encourage the participation of non-governmental entities (private sector) in space activities. While the *Draft Space Activities Bill* did not address the ownership of space resources, the *Indian Space Policy-2023* encourages non-governmental entities to "engage in the commercial recovery of an asteroid resource or a space resource" (Ud Din, 2021). The Policy states that "Any NGE engaged in such process shall be entitled to possess, own, transport, use, and sell any such asteroid resource or space resource obtained in accordance with applicable law, including the international obligations of India".

Thus, India's space policy envisions the extraction, possession, and commercial use of space resources, but makes these activities subject to applicable laws that include international legal frameworks. As already stated above, there is a regulatory gap in the Outer Space Treaty relating to ownership of space resources, which introduces a degree of ambiguity in India's policy on space resources. However, India's decision to join the Artemis Accords has introduced a significant change to this situation.

6.4.2 India and the Artemis Accords

India joined the Artemis Accords on June 21, 2023 (O'Shea, 2023). While signing the Artemis Accords India India's Ambassador to the United States emphasised the importance of a rule-based order for space exploration. As already mentioned, the Artemis Accords are concerning as they could potentially undermine a multilateral approach towards regulating outer space affairs, particularly through the adoption of the definitive interpretation on the ownership of space resources outside the consensus-driven framework of the UN system (Ud Din, 2022). In light of India's longstanding role as a leader of the aspirations of the Global South, how should India's joining of the Artemis Accords be interpreted?

One perspective is to adopt a critical view of India's joining of the Artemis Accords, interpreting it as a form of co-option into the US-led framework for space exploration, at the cost of a consensus-based multilateral approach towards managing outer space affairs. Alternatively, India's relationship with the Artemis Accords can be visualised from a different perspective. In the current political landscape relating to outer space exploration, India is uniquely positioned. It is the only significant spacefaring power that, on the one hand, aspires to represent the aspirations of the Global South and, on the other hand, does not have a confrontational space rivalry with the USA or the Russia-China bloc.

The distinct positioning allows India to leverage its role to push for revitalisation of a multilateral approach towards the management of space resources. As already stated earlier, it is the process through which decisions relating to the ownership of space resources are made that is of crucial significance, rather than the actual outcome. A multilateral approach within the UN system could steer space exploration in the direction of cooperation and healthy competition, rather than confrontation. India could play a pivotal role in establishing a multilateral approach towards the management of space resources and contribute towards establishing a collaborative framework for space exploration.

6.5 Conclusion

The commercial use of space resources is no longer confined to the realms of science fiction. Technological advancements have made use and exploitation of space resources feasible in the foreseeable future. However, the technological costs associated with the exploitation of space render them scarce and extremely valuable. At present, there appears to be a growing consensus within the international community that the commercial use of space resources is permissible under the Outer Space Treaty. Nevertheless, unlike the national space policies of several countries, the international space law does not explicitly address the ownership of space resources. The Artemis Accords, established in 2020, take a definitive stance on ownership and commercial use of space resources. In contrast, China and Russia have initiated a collaborative effort to launch a competing framework for lunar exploration, aimed at challenging the Artemis Accords. These developments indicate that the evolution of international space law, particularly concerning the ownership and use of space resources, is developing in a fragmented manner. However, one positive aspect is that China has demonstrated a willingness to advance international law on space resources through the COPUOS. Furthermore, the draft principles being developed at COPUOS represent a positive and constructive initiative.

More important than the specific legal framework governing the use of space resources is the process through which it is established. It is crucial to return to a multilateral approach within the ambit of the UN to advocate for a consensus-based regulatory framework for space resources. In this context, India is uniquely situated. With the successful landing of Chandrayaan-3, India has cemented its status as a major space power. India has also joined the Artemis Accords. As a space power that aspires to represent the interests of the Global South actors while maintaining a non-adversarial relationship with the US and China-Russia bloc, India is well placed to shape the future trajectory of international space law. The policies adopted by emerging powers like India could significantly influence whether future space exploration proceeds along cooperation or confrontation. In the years ahead, more attention needs to be focused on developments within COPUOUS. Notably, the draft principles being developed at COPUOS do not explicitly conflict with the provisions of Artemis Accords regarding the use of space resources. It is essential that emerging powers, such as China and India, actively support the work at COPUOS to help establish it as the primary forum for framing rules on space resources.

References

Christol, C. Q. (1980). The common heritage of mankind provision in the 1979 Agreement Governing the Activities of States on the Moon and Other Celestial Bodies. *International Lawyer, 14*(3), 429–430.

COPUOS. (2025, March 27). *Initial draft set of recommended principles for space resource activities*. Committee on the Peaceful Uses of Outer Space. https://documents.un.org/doc/undoc/gen/v25/021/86/pdf/v2502186.pdf

Goldie, L. F. E. (1983). A note on some diverse meanings of "the common heritage of mankind". *Syracuse Journal of International Law and Commerce, 10*(2), 69–105.

ISRO. (2017, February 15). *PSLV-C37 successfully launches 104 satellites in a single flight*. https://www.isro.gov.in/SuccessfullyLaunches104.html

ISRO. (2024a). *Mars Orbiter Mission*. https://www.isro.gov.in/MarsOrbiterMissionSpacecraft.html

ISRO. (2024b). *Indian Space Policy-2023*. https://www.isro.gov.in/media_isro/pdf/IndianSpacePolicy2023.pdf

Ji, E., Cerny, M. B., & Piliero, R. J. (2020, September 17). What does China think about NASA's Artemis Accords? *The Diplomat*. https://thediplomat.com/2020/09/what-does-china-think-about-nasas-artemis-accords/

Jones, A. (2024, March 6). China outlines position on use of space resources. *SpaceNews*. https://spacenews.com/china-outlines-position-on-use-of-space-resources/.

Lele, A., & Gopalakrishnan, V. (2020, October 24). *Artemis Accords: Unilateralization in space*. Society for the Study of Peace and Conflict. https://sspconline.org/opinion-analysis/artemis-accords-unilateralization-space-sat-10242020

Luxembourg Space Agency. (2019, November 18). *Spaceresources.lu initiative*. https://space-agency.public.lu/en/space-resources/the-initiative.html

Melchin, G. (2015). You can't take the sky from me: A Gramscian interpretation of the common heritage of mankind principle in space law. *Dalhousie Journal of Legal Studies, 24*, 141.

Merges, R. P., & Reynolds, G. H. (1997). Space resources, common property, and the collective action problem. *NYU Environmental Law Journal, 6*, 105.

NASA. (2020, October 13). *The principles for cooperation in the civil exploration and use of the Moon, Mars, comets, and asteroids for peaceful purposes*, Section 10(2). https://www.nasa.gov/wp-content/uploads/2022/11/Artemis-Accords-signed-13Oct2020.pdf

NASA. (2024, October 18). *The Artemis Accords*. https://www.nasa.gov/artemis-accords/
Noyes, J. E. (2012). The common heritage of mankind: Past, present, and future. *Denver Journal of International Law and Policy, 40*(1–3), 447.
O'Shea, C. A. (2023, June 23). NASA welcomes India as 27th Artemis Accords signatory. *NASA*. https://www.nasa.gov/news-release/nasa-welcomes-india-as-27th-artemis-accords-signatory/
Pandey, G. (2023, August 23). Chandrayaan-3: India makes historic landing near Moon's south pole. *BBC News*. https://www.bbc.com/news/world-asia-india-66594520
Thomas, J. (2005). Privatization of space ventures: Proposing a proven regulatory theory for future extraterrestrial appropriation. *Brigham Young University International Law & Management Review, 1*(1), 191.
Tingley, B. (2023, August 24). *India's successful Chandrayaan-3 Moon landing "belongs to all of humanity," Prime Minister Modi says*. Space.com. https://www.space.com/india-chandrayaan-3-moon-landing-prime-minister-modi-all-humanity
Tjandra, J. (2021). The fragmentation of property rights in the law of outer space. *Air and Space Law, 46*(3), 373–391.
UAE Space Agency. (2023). *Space resources regulation: Regulatory framework on space activities of the United Arab Emirates*, Art. 7. https://space.gov.ae/en/policy-and-regulations
Ud Din, A. (2021). India's quest for a national space law and the missing piece of possessory rights. *India Quarterly, 77*(4), 642.
Ud Din, A. (2022). The Artemis Accords: The end of multilateralism in the management of outer space? *Astropolitics, 20*(2–3), 135.
United Nations. (1967a). *Treaty on principles governing the activities of states in the exploration and use of outer space, including the moon and other celestial bodies*, 610 U.N.T.S. 205.
United Nations. (1967b, November 1). UN Doc P A/C.1/PV.1515, para 58.
United Nations Committee on the Peaceful Uses of Outer Space. (1979). *Summary records of the one hundred and eleventh to the one hundred and thirty-first meetings, from June 09 to July 04, 1969* (UN Doc A/7285). https://www.unoosa.org/pdf/gadocs/A_7285E.pdf
United Nations Office for Outer Space Affairs. (2024). Submission by the Delegation of China to the Working Group on Legal Aspects of Space Resource Activitiesof the Legal Subcommittee of the Committee on the Peaceful Uses of Outer Space. United Nations Office for Outer Space Affairs. https://www.unoosa.org/documents/pdf/copuos/lsc/space-resources/LSC2024/English_Chinas_submission_to_the_working_group_on_space_resources.pdf
United States. (2010). *National space policy of the United States of America*. NASA. https://www.nasa.gov/sites/default/files/national_space_policy_6-28-10.pdf
U.S. Commercial Space Launch Competitiveness Act. (2015). *U.S. Commercial Space Launch Competitiveness Act*, Section 513.02.a. https://www.govinfo.gov/content/pkg/BILLS-114hr2262enr/pdf/BILLS-114hr2262enr.pdf
Wall, M. (2022, September 3). *Not just Artemis: China and Russia plan to put boots on the Moon, too*. Space.com. https://www.space.com/china-russia-moon-base-ilrs
Wu, X. (2023). The International Lunar Research Station: China's new era of space cooperation and its new role in the space legal order. *Space Policy, 65*, 101537. https://doi.org/10.1016/j.spacepol.2022.101537
Yarlagadda, S. (2024, October 24). *Economics of the stars: The future of asteroid mining and the global economy*. Harvard International Review. https://hir.harvard.edu/economics-of-the-stars/
Zhao, Y. (2004). An international space authority: A governance model for a space commercialization regime. *Journal of Space Law, 30*(2), 283.

Chapter 7
Artemis Accords as Evolutive Law-Making: Lunar Space Mining and the Rise of Space Militarisation

Adithya Variath

Abstract The new age of technological developments continues to bring humanity much closer to Moon surface mining. Tapping the lunar economy could provide humanity with the potential to expand its presence on the Moon, as well as further into the solar system. In this context, the Artemis Accords and the geopolitics around it are shaping new complexities of lunar resource extraction. While exploiting the lunar economy is the primary objective of the Artemis Accords, the non-binding principles apply to civil activities in outer space, and all activities that may take place on the Moon, Mars, comets, and asteroids, including their surfaces and sub-surfaces, as well as in orbit of the Moon or Mars, in the Lagrangian points for the Earth-Moon system, and in transit between these celestial bodies and locations. The United States drafted the Accords is building consensus, and as of January 2025, the number of signatories has grown to 53. The Artemis Accords express a soft obligation to 'reinforce and implement the Outer Space Treaty, the Registration Convention, the Agreement on the Rescue and Return of Astronauts' and 'other norms of behaviour that NASA and its partners have supported'. However, some of the principles raise concerns about its consistencies with the fundamentals of space law as it introduce concepts such as safety zones, resource extraction and use and interoperability. It creates new challenges that raise questions about sovereignty, commercial rights, sustainability in space and potential military use of the lunar environment, as nations may leverage lunar activities for strategic dominance. This chapter explores the Artemis Accords as a new form of evolutive lawmaking redesigning global space governance through unipolarism masquerading as multilateralism. The chapter primarily focuses on how the Accords can foster competitive dynamics among space-

Adithya Variath, "Artemis Accords as Evolutive Law-Making: Lunar Space Mining and the Rise of Space Militarisation", in Sandeepa Bhat & Adithya Variath, *Artemis Accords and Resource Mining in Outer Space* (Issues in Space: Springer) Forthcoming, 2025.

A. Variath (✉)
Jindal Global Law School, O.P. Jindal Global University, Sonipat, India
e-mail: adithya.variath@jgu.edu.in

S. Bhat B., A. Variath (eds.), *Artemis Accords and Resource Mining in Outer Space*, Issues in Space, https://doi.org/10.1007/978-981-95-0152-6_7

faring nations, which may accelerate the development of dual-use technologies under the facade of peaceful exploration. By placing the Accords within the broader context of astropolitical diplomacy, this chapter explores the future of lunar space mining and how the Accords will shape the future norms of militarisation and commercial exploitation of the lunar environment.

Keywords Lunar mining · Space militarisation · Artemis accords · NATO · Space force · Space weapons

7.1 Introduction

In 1972, Harrison Schmitt, a geologist with NASA, became the last person to walk on the moon (Benaroya, 2010). Space-faring states are still looking into ways to return to the lunar surface more than fifty years later. Schmitt's groundbreaking book, *Return to the Moon*, published in 2006, provides a roadmap for the hunt for Earth's future energy sources (Schmitt, 2007). He presents a strong scientific argument in his book for going back to the Moon, which has undeveloped minerals worth billions of dollars, including Helium-3 (3He). One of the most significant sources of sustainable energy could be helium-3. It might be the perfect fuel for fusion reactors because it is a non-radioactive isotope. Helium-3's potential for use in nuclear fusion reactors was emphasised in a NASA report from 1988. "Compared to current nuclear power, it offers several advantages as an abundant, low-carbon energy and no nuclear waste technique," according to scientists (NASA, 1988). The difficulty lies in capturing and limiting this energy to make it a practical source in the near future, even though many of these concepts are still in the experimental stage (Variath, 2024a).

Both opportunity and necessity motivate the exploration and use of space resources. Leveraging in-situ resources is essential for lowering costs, minimising reliance on Earth, and permitting long-term living as mankind aims to establish a sustainable presence in space. Rare minerals and metals found on celestial worlds have the potential to transform Earthly businesses, while materials like water can be transformed into rocket fuel and life support systems (Dallas et al., 2020; Sivolella, 2019). NASA has proven that there are ice deposits inside deep craters close to the lunar south pole, and water is one of the most valuable minerals in space (Sivolella, 2019).

In addition to providing astronauts with a means of subsistence, water taken from the moon might be transformed into hydrogen and oxygen, which together provide a potent rocket propellant. This would then make it possible to explore farther-off places, such as mineral-rich asteroids in the asteroid belt between Mars and Jupiter, and it might open up a world of economic and social benefits for the participating countries (Leterre, 2024). However, several difficulties and unknowns accompany the potential of space resources. Due to the need for previously unheard-of breakthroughs in the extraction and processing of these resources, there are still substantial scientific, technological, and financial obstacles. Furthermore, the governance and legal structures governing the use of resources are still developing. The potential of the moon and other celestial bodies as natural resources has been acknowledged by powerful space-faring nations such as the United States (Hubbard et al., 2024). Spacecraft from China, India, and Israel have visited the moon in recent years, and a South Korean orbiter is in the works.

In this regard, NASA's "Artemis I" mission's successful launch marks a first step towards establishing a long-term human presence on the moon and potentially exploiting enormous amounts of incredibly valuable commodities there and elsewhere in space (NASA, 2020b). However, spacefaring governments should set more specific guidelines about who will receive this bounty and how before the mining starts. In support of its Artemis program, NASA created a new normative structure, Artemis Accords (hereafter 'Accords'), described as "a shared vision for principles, grounded in the Outer Space Treaty of 1967, to create a safe and transparent environment which facilitates exploration, science, and commercial activities for all of humanity to enjoy" (NASA, 2020a). On 13 October 2020, seven countries—Australia, Canada, Italy, Japan, Luxembourg, the United Arab Emirates and the United Kingdom—signed onto the United States-led Accords. As of 20 February 2025, the number has increased to 53. On 21 January 2025, Finland became the 53rd nation to sign the accords (NASA, 2020b).

The Outer Space Treaty (hereafter 'OST') and its interpretations serve as the foundation for the current regulatory regime for all space activities, independent of multinational negotiations such as the Artemis Accords (Aoki, 2024). However, the OST is not clearly defined and dates from a period when nation-states were only beginning to engage in space activities (UNIDR, 2007). Furthermore, neither the United States nor any other significant spacefaring nation have ratified the 1979 Moon Agreement, which stipulates that neither the moon nor its resources "shall become a property of any state, international intergovernmental or nongovernmental organisation, national organisation, or nongovernmental entity, or of any natural person" (Delgado-López, 2015). Several nations have separately created national plans to make use of these technical developments and space resources. According to the 2015 U.S. Commercial Space Launch Competitiveness Act, US citizens are "entitled to any asteroid resource or space resource obtained in accordance with applicable law, including the international obligations of the United States" (*U.S. Commercial Space Launch Competitiveness Act of 2015: report of the Committee on Commerce, Science and Transportation on S. 1297*, 2015).

In order to "ensure that there is a norm of behaviour" while collecting and using such resources, NASA recently chose four commercial businesses to gather lunar samples and sell them to NASA, helping to set a precedent in this regard. India, Japan, the United Arab Emirates, and Luxembourg have all enacted unilateral laws and policies pertaining to the mining of space resources (Salmeri, 2023). China and Russia have also established aspirations to work together on moon missions and space mining initiatives (Jakhu et al., 2016a, 2016b). The idea of safety zones outlined in the Accords has drawn criticism from China, while the chief of Russia's space program has said that the Accords are now too "US-centric" to agree upon (Deplano, 2021).

The Accords cover all civil activities in outer space, including those that may occur on the Moon, Mars, comets, and asteroids, including their surfaces and subsurfaces, as well as in orbit of the Moon or Mars, in the Lagrangian points for the Earth-Moon system, and transit between these celestial bodies and locations, even though the Moon is currently the main mission objective. The Artemis Accords offer a shared set of guidelines to improve the governance of the civil exploration and use of space while numerous nations and private enterprises carry out missions and operations around the Moon. The Outer Space Treaty, the Registration Convention, the Rescue and Return Agreement, and best practices and standards of responsible behaviour for civil space research and utilisation are all reaffirmed by the Artemis Accords. The Accords' conformity with other international legal documents has drawn criticism.

To "improve the governance of the civil exploration and use of outer space" and "increase the safety of operations, reduce uncertainty, and promote the sustainable and beneficial use of space for all humankind," the Artemis Accords include ten principles. It is understood that these guidelines are not meant to be a comprehensive statement of all the guidelines that ought to govern actions carried out in accordance with the Accords. It is anticipated that participants would sign particular agreements that embrace the Accords' tenets and specify additional requirements, including liability, intellectual property ownership and usage, and the interchange of commodities and data (Din, 2022; Kostenko, 2021; Tronchetti & Liu, 2021).

7.2 Problematising Artemis Accords and Conceptualising Space Weapons

7.2.1 Inconsistencies of Artemis Accords

The principles of the Accords are expressed to 'represent a political commitment to the principles described herein, many of which provide for operational implementation of important obligations contained in the Outer Space Treaty and other instruments'(NASA, 2020a). The Preamble to the Accords affirms the importance of compliance with four of the United Nations Space Treaties, namely the Treaty on Principles Governing the Activities of States in the Exploration and Use of Outer Space, Including the Moon and Other Celestial Bodies; the Agreement on the Rescue of Astronauts, the Return of Astronauts and the Return of Objects Launched into Outer Space; the Convention on International Liability for Damage Caused by Space Objects; and the Convention on Registration of Objects Launched into Outer Space. The Preamble to the Accords also affirms 'the benefits of coordination via multilateral forums, such as the UN Committee on the Peaceful Uses of Outer

Space, to further efforts toward a global consensus on critical issues regarding space exploration and use', the fifth UN space treaty, the Agreement Governing the Activities of States on the Moon and Other Celestial Bodies (Moon Agreement), is absent from this list. The US is also not a party to the Moon Agreement, which has only 17 signatories. None of the traditional major space-faring States are ratified States Parties to the Moon Agreement, and it is apparent that the Moon Agreement is not reflective of customary international law. This view has been further affirmed most recently by the US in the context of the Artemis Accords.

The 2020 US 'Executive Order on Encouraging International Support for the Recovery and Use of Space Resources' clearly stated: "The United States does not consider the Moon Agreement to be an effective or necessary instrument to guide nation states regarding the promotion of commercial participation in the long-term exploration, scientific discovery, and use of the Moon, Mars, or other celestial bodies. Accordingly, the Secretary of State shall object to any attempt by any other state or international organisation to treat the Moon Agreement as reflecting or otherwise expressing customary international law" (Trump, 2020). The Executive Order expressly restates the US's stance that it will oppose "any attempt to treat the 1979 Moon Agreement as expressing customary international law," even though the White House statement that goes with it reiterates that the Order is primarily directed at US policy regarding resource recovery and utilisation in space. While some of the Accords' sections reflect ideas or standards being discussed in ongoing working groups, others aim to go beyond the current consensus positions on space uses, and still others are essentially a restatement of the Outer Space Treaty's existing space law principles. It has been claimed that the Accords propagate and uphold US conceptions of international space law in this way. These discrepancies mostly concern the interplay between sections 10 and 11 of the Accords, which permit in situ space resource utilisation and the establishment of "safety zones," and Article II of the Outer Space Treaty, which forbids States from asserting sovereignty in outer space "by means of use or occupation" (Lee, 2004; Tronchetti, 2008).

7.2.2 The Curious Case of Art. IV and Space as an Operational Domain of Warfare

Since the launch of Sputnik in 1957, military use has been a key feature of space activities and the space race. Over the last two decades, space-faring states have developed a wide range of conventional and non-conventional weapons (Variath, 2024c). Today, outer space, valued at $1.8 trillion, is becoming a highly contested zone (*Space: The $1.8 trillion opportunity for global economic growth*, 2024). With the growing value of the space economy, the element of weaponisation is essential to protect critical space assets from a pre-emptive attack. However, the larger threat is the growing fear of space warfare. The recent entrants to this domain have been modern kinetic and non-kinetic weapons (Bongers & Torres, 2024). The United

States, Russia, China and India have developed and tested these weapons. Israel is thought to have Anti-Satellite (ASAT) weapons capability, but it has not tested it yet (Rajagopalan, 2022). There is no treaty explicitly banning these weapons. However, there has been global momentum to regulate these weapons.

In 2022, the United States administration said the US would commit "not to conduct destructive, direct-ascent missile testing and that seeks to establish this as a new international norm for responsible behaviour in space" (Harris, 2022). However, China and Russia have blocked all the moves towards de-weaponisation. The geopolitics of space weapons indicate that there is no consensus when it comes to the regulation of weaponisation. At the global level, the United States, Russia, China, the United Kingdom, and India have integrated space weapons into their military apparatus. In 2019, NATO adopted a new Space Policy, which declared outer space the fifth operational domain to protect its critical infrastructure and to strengthen its resilience in space. The policy states that NATO's approach to space will remain fully in line with international law (NATO, 2019). Globally, defence spending has risen to 45% of government space budgets (Variath, 2024b). While the military use of space is evolving, the space law framework is stagnating.

The current era of lawmaking in outer space is marked by the decline of the 'traditional' treaty method. Although the law-making has stopped after the Moon Agreement in 1979, in the context of space weapons, technological developments have been on the rise at a very rapid pace. The need to prevent the evil effects of space militarisation was felt immediately after the beginning of the space race in the 1950s (Jasani & Stockholm International Peace Research, 1987). In response to this need, the UN General Assembly passed its first resolution on outer space to the domain of peaceful use after the launch of Sputnik. It recognised the common interest of mankind in outer space and stressed that the use of outer space should be for 'peaceful purposes only' (Pope, 2021). However, the absence of a definition of 'peaceful uses' has curtailed all efforts to prevent the evils of space militarisation.

The 1967 Outer Space Treaty, often referred to as the Magna Carta of space law, has proved to be powerless to regulate modern space weapons. Article IV of the OST states: "States Parties to the Treaty undertake not to place in orbit around the earth any objects carrying nuclear weapons or any other kinds of weapons of mass destruction, install such weapons on celestial bodies, or station such weapons in outer space in any other manner." The literal interpretation of Article IV indicates that it prohibits the "placement" or "installation" of weapons in orbit while many modern weapons including the ASAT does not involve either (Bhat & Gupta, 2023). Article IV only prohibits "weapons of mass destruction" and "nuclear weapons" and excludes the proliferation of other types of weapons or emerging technologies like anti-satellite systems, dual-use technologies and directed-energy weapons.

In 2019, the United States created *Space Force* to develop space-based force support technologies and space military operations. Article IV also states that "the moon and other celestial bodies shall be used by all States Parties to the Treaty exclusively for peaceful purposes." The United States has long argued that its military operations are an effort to maintain the peaceful use of outer space. Even the Space Force's June 2020 doctrine document states that "in keeping with international law, the United States acknowledges that the use of space is for peaceful purposes" (*Space Power: Doctrine for Space Forces*, 2020). This is a trend that has been built on the United States' approach to use space for military activities.

In 1959, the United States created the "US Policy in Outer Space" as an effort to reinterpret America's early advocacy for using space for exclusively peaceful purposes (Council, 1959). It was at the backdrop of the United Nations General Assembly debating the "Declaration of Legal Principles". By then, the United States had largely solidified its reinterpretation of the use of outer space for peaceful purposes. Senator Albert Gore, while addressing the United Nations First Committee on the subject of international cooperation in space said: "it is the view of the United States that outer space should be used only for peaceful—that is, non-aggressive and beneficial—purposes. Until [general disarmament] is achieved, the test of any space activity must not be whether it is military or non-military, but whether or not it is consistent with the United Nations Charter and other obligations of law" (Gore, 1962). The American shift from the "exclusively" peaceful use to the "non-aggressive" peaceful use continues to shape how United States approach space. The United States, adopting this formulation, argued that its plans of Strategic Defense Initiative (Star Wars Program), Anti-ballistic Missile, Anti-satellite weapons, nuclear defence, jamming, spoofing and cyber-attacks, etc., were all defensive and hence 'peaceful'.

There is also a concern about whether international humanitarian law can be extended to outer space so that principles of distinction and proportionality can be used to assess the "legality" of weapons. The period from 1984 to 2020 can be called a period of space law 'winter' wherein there has been no treaty-making at the UN level. However, there have been multiple soft law mechanisms developed by various bodies. A fundamental shift in this regard has been the NASA-drafted Artemis Accords building a new form of consensus at the global level.

The most common approach to creating international rules governing space is "soft law," which has grown in popularity over time (Mercer & Queen Elizabeth, 2009). Soft law tools are morally and normatively valuable but do not impose legally binding responsibilities on states (Marboe, 2012). In addition to actively assisting in the progressive construction and consolidation of international law standards, they can significantly influence states' behaviour. The idea that the legal value of soft law should be determined by the parties' intention rather than the name given to a legal document is put forth by a substantial approach to soft law.

Resolutions passed by the UN General Assembly have tried to close the gaps left by the demise of treaties by allowing for the establishment and discussion of space law without requiring states to ratify legally binding agreements (Bhat, 2024). The failure of China and Russia's joint submissions to the UN Conference on Disarmament in 2008 and 2014 regarding the Draft Treaty on the Prevention of the Placement of Weapons in Our Space and of the Threat of Force against Outer Space Objects demonstrates this preference for soft law over legally binding multilateral treaties (Economic & Security Review Commission, 2017; Kathryn, 2022). General Assembly Resolutions outlining the pertinent principles have been adopted instead of the proposed treaty, which has had limited success in establishing a legally binding commitment.

7.3 Artemis Accords, Space Weapons and Lunar Space Mining

7.3.1 Peaceful Purposes and Space Resources

Section 10 of the Accords deals with 'Space Resources' and states that the Signatories acknowledge that the use of space resources can help humanity by offering vital assistance for operations that are sustainable and safe. In accordance with the Outer Space Treaty and support of safe and sustainable space activities, the Signatories stress that the extraction and use of space resources, including any recovery from the surface or subsurface of the Moon, Mars, comets, or asteroids, can be carried out (Byers & Boley, 2023). According to Article II of the Outer Space Treaty, the signatories state that contracts and other legal documents pertaining to space resources shall be in accordance with the Treaty and that the extraction of space resources does not automatically amount to national appropriation. The Signatories intend to use their experience under the Accords to contribute to multilateral efforts to further develop international practices and rules applicable to the extraction and utilisation of space resources, including through ongoing efforts at the UNCOPUOS (de Zwart & Lisk, 2024).

China and Russia, two rival spacefaring nations, have rejected the Accords, with Russia claiming they are too US-centric. According to Chinese experts, the US is approaching this new age of space competition with a Cold War mindset (Deplano, 2021). Given the ongoing rivalry between the US and China and the current challenges in spacefaring states achieving a consensus at the UN, this response is hardly surprising. The US can use the Accords to regulate and influence how international space law develops in the future. Artemis Accords will cause fragmentation and competition and a breakdown of the order created by the Outer Space Treaty. Russia criticised President Trump's Executive Order on Space Resources and the Artemis Accords, accusing the US of colonising and privatising space ("Russia Compares Trump's Space Mining Order to Colonialism," 2022). Russia and China believe that the Accords aim to build a community of interests led by the US in the field of lunar and deep space exploration, as well as to establish its dominant position in space resources exploitation and recovery, and, finally, to consolidate its leading position in space.

Naturally, only states have the power to create international law through the draughting of multilateral treaties and the establishment of new customary international law through widespread and consistent state practice. The accords mirror the structure of current international space law and provide a clear interpretation of controversial and ambiguous problems. Therefore, the Artemis Accords offer a different and thus opposing perspective on a number of important international space law problems that could jeopardise multilateral approaches to space activity and heighten the likelihood of conflict. Diplomacy, the development of soft law, and behavioural standards all function beneath international law (Riordan et al., 2023). Notably, there are other multistate initiatives aiming to place people on the Moon besides Artemis. In 2021, China and Russia invited other nations to participate in the International Lunar Research Station (hereafter 'ILRS'), their plans to build a permanent human outpost on the moon (Wu, 2023). Similar to Artemis, the ILRS will include human and scientific research in addition to extractive and utilisation activities (Xu & Ou, 2023).

Section 3 on Peaceful Purposes states 'that cooperative activities under these Accords should be exclusively for peaceful purposes and in accordance with relevant international law'. This requirement echoes the wording of the Outer Space Treaty, in which the term 'peaceful purposes' appears four times: twice in the preamble and twice in Article IV. Article IV articulates the agreement by States Parties to the Outer Space Treaty to refrain from placing nuclear weapons or other weapons of mass destruction in orbit around the Earth, on celestial bodies or stationing 'such weapons in space in any other manner' (de Zwart & Lisk, 2024). The second paragraph of this Article states: "The Moon and other celestial bodies shall be used by all States Parties to the Treaty exclusively for peaceful purposes. The establishment of military bases, installations and fortifications, the testing of any type of weapons and the conduct of military manoeuvres on celestial bodies shall be forbidden. The use of military personnel for scientific research or for any other peaceful purposes shall not be prohibited. The use of any equipment or facility necessary for the peaceful exploration of the Moon and other celestial bodies shall also not be prohibited." The term "peaceful purposes," which was used in the first UN General Assembly resolution is a fundamental and evident foundational principle for space governance. It remains to be seen how will peaceful behaviour play out with the involvement of civilian space agencies and commercial partners.

7.3.1.1 Safety Zones

Section 11 presents the most critical issue in the debate—deconfliction of space activities-including the creation and recognition of 'safety zones'. The Artemis Accords' Section 11 addresses the matter of space activity deconfliction. Its paragraphs 3 and 4 make reference to two principles found in Article IX of the Outer Space Treaty: "the duty to refrain from detrimental interference with the activities of other States Parties to the treaty" and "the principle that States should conduct their activities in outer space with due regard to the corresponding interests of the other States Parties" (de Zwart & Lisk, 2024). Section 11, paragraph 5 of the Artemis Accords mandates that the signatories establish a "safety zone" to prevent potentially detrimental interference and give each other the "necessary information regarding location and nature" of their actions in order to operationalise such principles. Safety zones are intended to be transitory and have two functions. First, they want to make scientific research easier. Second, they serve as a way to facilitate "the safe and efficient extraction and utilisation of space resources in support of sustainable space exploration and other operations," building on the provisions of Section 10 of the Artemis Accords.

To implement these principles, Signatories agree to "contribute to multilateral efforts to further develop international practices, criteria, and rules applicable to the definition and determination of safety zones and harmful interference" by drawing on their experience under the Accords. The fact that these safety zones are stated to be carrying out duties under the Outer Space Treaty, such as avoiding harmful interference, is noteworthy. Signatories must designate specific areas as "safety zones" and notify others of their presence in order to reduce the possibility of detrimental interference brought on by a particular activity or "anomalous event" (Takeuchi, 2022).

The section states that the "size and scope of the safety zone, as well as the notice and coordination, should reflect the nature of the operations being conducted and the environment that such operations are conducted in" and "leveraging commonly accepted scientific and engineering principles," without giving any specific measurements or dimensions for a safety zone (Mallowan et al., 2021). Furthermore, it is not clear how long these zones are supposed to last. It is intended to be transitory and will disappear when the relevant procedure stops (Takeuchi, 2022). According to the Artemis Accords, "the nature and existence of safety zones is expected to change over time reflecting the status of the relevant operation," acknowledging the controversial nature of the safety zones concept as well as a growing understanding of the space environment. The operational Signatory should modify the size and scope of the related safety zone as necessary if the nature of an operation changes (Gilbert, 2023).

The Signatories commit to respect all other provisions of the Outer Space Treaty and the principle of free access to all areas of celestial bodies in their use of safety zones. This clause seems to walk a tightrope between a safety zone being just a question of safety and a safety zone turning into a de facto claim to a portion of space for "safety" reasons (de Zwart & Lisk, 2024). In the interim, Signatories are required by Section 11 of the Accords "to adjust their usage of safety zones over time based on mutual experiences and consultations with each other and the international community." International space law does not explicitly recognise safety zones, despite the fact that the idea has been acknowledged as having the ability to effectively implement the necessity of preventing collisions and conflicts. There is a 200-kilometre "keep out" zone around the International Space Station (NASA, 2019).

Safety zones run the possibility of being used as a cover for militarisation and territorial control, even if they are presented as tools to improve collaboration and lessen conflict. The creation of safety zones may resemble territorial claims even when they are not intended to grant sovereignty, particularly if they are implemented unilaterally or without international agreement. States may use technology like active security mechanisms, cyber defences, or surveillance systems to safeguard activities within safety zones. The distinction between militarised and peaceful activities may become hazier as a result of these acts, which could increase tensions. The establishment of safety zones by several countries or coalitions may result in rivalry for areas with abundant resources, raising the possibility of space wars. Although the Accords promote private sector involvement, the absence of explicit rules governing their security operations may result in militarised reactions masquerading as safeguarding business operations.

Such safety measures shall not impede the free access, in accordance with international law, to any area of outer space by personnel, vehicles and equipment of another operator. In accordance with the area-based safety measure, a State or international organisation may restrict access for a limited time, provided that timely public notice has been given setting out the reasons for such restriction. The Artemis Accords' safety zones are essential for safeguarding large expenditures made on space operations and exploration (McKeown et al., 2022). Massive financial and technological resources are needed for space activities, such as habitat development, lunar mining, and research projects. Long-term plans for space development could be derailed, and these investments are put at risk by any disruption, whether deliberate or unintentional. By defining areas where activities can continue without outside intervention, safety zones act as protective buffers, assisting in shielding operations from detrimental interference (Vertadier & Kate Schingler, 2023). This guarantee is especially important for building trust with private-sector stakeholders, whose participation is critical to expanding space's commercial and scientific promise.

The problem is made more difficult by the uncertainty surrounding the dual-use nature of many technologies. There could be an arms race in space if systems put in place to guard safety zones are also used offensively. The presence of private organisations, which may implement their security measures without sufficient international monitoring, adds to this difficulty. In violation of the non-appropriation principle of the Outer Space Treaty, such acts run the risk of intensifying the sense of territorial authority within safety zones.

7.4 Artemis Accords as Evolutive Law-Making

The Accords cannot be classified as "hard law" because they do not establish legally binding duties. Accords are a part of the expanding field of soft law, which is addressing the gaps in space law and governance. However, Soft law does not imply a "no" law (Meyer, 2010). International law is ultimately created by states, and the Accords offer an intriguing case study for the ongoing evolution of law-making at the global level. This can be called as a new era of space law-making that is evolutionary, wherein consensus is built through soft law mechanisms (Rachel, 2021). The Accords can be a new normative tool to understand the new form of governance, even though they are only political agreements. The Accords are not the first time a few states have agreed to a space mission. The Intergovernmental Agreement, which outlines the duties of each State and its crew, established and governs the International Space Station (NASA, 2020c). The Intergovernmental Agreement imposed legally binding responsibilities on its parties, but the Accords established non-binding standards of conduct for its Signatories. This is the main distinction between the two agreements.

The Accords were mostly drafted by the United States, in contrast to earlier soft law instruments that were created by UNCOPUOS and presented to all States before the UN General Assembly. The Accords represent one State's attempts to start interpreting or maybe changing the parameters of outer space law. Sections on safety zones, space resources, and heritage all state that "the Signatories intend to use their experience under the Accords to contribute to multilateral efforts to further develop international practices" and regulations, which raises the possibility that the Accords encourage multilateralism.

The Russia and China initiative International Lunar Research Station, although a project currently just between the two States, state partners can join it through negotiations with China and Russia, according to the *Guide for Partnership*, which was released in June 2021 (*International Lunar Research Station (ILRS) Guide for Partnership*, 2021). Many the content of these legal documents is still unknown, and China and Russia may attempt to add their interpretations of the Outer Space Treaty. The State Corporation for Space Activities in Russia has previously condemned the United States' stance on space mining, claiming that "attempts to expropriate outer space and aggressive plans to take over other planets" discourage international cooperation. On the other hand, the commercial aerospace company Origin Space Co Ltd. launched a mining spacecraft prototype in April 2021, demonstrating China's increased acceptance of resource use (*China launches robot prototype capable of catching space debris with net*, 2021).

The Artemis Accords offer a possible new avenue for the development of space law, whereby powerful spacefaring nations attempt to employ their soft power to advance legal interpretations that serve their own interests. One could claim that the multilateral accords originated in this manner. The two proposals put up by the two leading space powers at the time, the US and the USSR, were essentially combined to create the Outer Space Treaty. The fact that major space countries are now advancing projects without wider international agreement, however, is noticeably different. This could prevent smaller, non-spacefaring countries from taking part in upcoming changes to the legislation governing outer space.

The development and deployment of space-based weapons under the pretence of peaceful usage or defence could be made easier by making use of ambiguities in the wording of soft law instruments like the Accords. Because there are no legally binding limitations, there is a risk that the militarisation of space may intensify, jeopardising both international security and the fundamental idea that space should be used for peaceful reasons (Bourbonnière & Lee, 2007; Khalid, 2021). The idea of "safety zones" as defined by the Accords, which are meant to avoid detrimental interference during space operations, is one such problem, as discussed in this chapter. Even while these areas are presented as instruments for operational coordination and transparency, their ambiguity could be used to support discriminatory actions or even covert militarisation. States may use defensive systems or dual-use technology that conflate military and civilian uses in the name of preserving security or safeguarding national resources. While encouraging creativity and adaptability, soft law's intrinsic flexibility also allows governments to evade more stringent international inspection, raising the possibility of a militarised space environment.

Therefore, even while the Accords are a new start for space governance, their shortcomings show how urgently strong, legally enforceable international accords are needed to stop the weaponisation of space.

7.5 Conclusion

The Artemis Accords represent a significant change in the governance of space, signifying a departure from universal, consensus-driven frameworks such as the Outer Space Treaty and a turn towards a more coalition-based approach to collaboration. The Accords highlight the changing nature of international space law, where norms are influenced by both multilateral accords and unilateral rewriting of rules and alliances, even as they seek to promote peaceful exploration and resource use on the Moon and beyond. As envisioned by the Accords, lunar space mining offers both possibilities and difficulties. Lunar mining promises tremendous economic gains and technical breakthroughs. However, the lack of a comprehensive international framework that addresses sustainability and benefit-sharing runs the risk of compromising the fundamental idea that space is the "province of all humankind." Furthermore, the peaceful use of space that was intended in previous treaties is in danger of being undermined by the strategic ramifications of space resource exploitation as well as increasing militarisation.

Maintaining space as a conflict-free environment is directly threatened by the emergence of space militarisation in the context of resource rivalry. Although the Artemis Accords place a strong emphasis on peaceful cooperation, a crucial weakness in the current system of governance is highlighted by the lack of safeguards, ambiguous provisions and geopolitical unilateralism (von der Dunk, 2021). Monitoring and reporting activity in safety zones in detail helps build trust and stop unilateral actions from turning into confrontations. Ambiguity and misuse can be decreased by defining precisely what 'harmful interference' is and what activities are allowed in safety zones. This emphasises the necessity of an evolving approach to space law that strikes a balance between sustainability, security, and technological advancement. Developing a legal and institutional framework that not only tackles the current issues of lunar space mining but also foresees and reduces the risks of geopolitical tensions and militarisation is crucial as humanity explores space for resources.

References

Aoki, S. (2024). Outer space treaty and fundamental principles. In *International space law in the new space era: Principles and challenges* (p. 0). Oxford University Press. https://doi.org/10.1093/9780198909415.003.0004

Benaroya, H. (2010). *Lunar settlements* (1st ed.). CRC Press. https://doi.org/10.1201/9781420083330

Bhat, B. S. (2024). An introduction to international space law. In *International space law in the new space era: Principles and challenges* (p. 0). Oxford University Press. https://doi.org/10.1093/9780198909415.003.0001

Bhat, B. S., & Gupta, R. (2023). Sui generis interpretation of key elements of space demilitarization for ensuring peaceful uses. *Indonesian Journal of International & Comparative Law, 10*(2), 113.

Bongers, A., & Torres, J. L. (2024). Star wars: Anti-satellite weapons and orbital debris. *Defence and Peace Economics, 35*(7), 826–845. https://doi.org/10.1080/10242694.2023.2208020

Bourbonnière, M., & Lee, R. J. (2007). Legality of the deployment of conventional weapons in earth orbit: Balancing space law and the law of armed conflict. *European Journal of International Law, 18*(5), 873–901. https://doi.org/10.1093/ejil/chm051

Byers, M., & Boley, A. (2023). Space mining. In M. Byers & A. Boley (Eds.), *Who owns outer space?: International law, astrophysics, and the sustainable development of space* (pp. 130–185). Cambridge University Press. https://doi.org/10.1017/9781108597135.006

China launches robot *prototype capable of catching space debris with net.* (2021). NY Post. Retrieved 20 December from https://nypost.com/2021/04/27/china-launches-robot-prototype-capable-of-catching-space-debris-with-net/

Council, N. A. a. S. (1959). *US Policy on Outer Space.* Marshall Institute: National Security Space Project, NSC 5918/1 Retrieved from **https://aerospace.csis.org/wp-content/uploads/2019/02/NSC-5918-US-Policy-on-Outer-Space.pdf**

Dallas, J. A., Raval, S., Gaitan, J. P. A., Saydam, S., & Dempster, A. G. (2020). Mining beyond earth for sustainable development: Will humanity benefit from resource extraction in outer space? *Acta Astronautica, 167*, 181–188. https://doi.org/10.1016/j.actaastro.2019.11.006

de Zwart, M., & Lisk, J. (2024). The artemis accords and subsequent developments. In *International space law in the new space era: Principles and challenges* (p. 0). Oxford University Press. https://doi.org/10.1093/9780198909415.003.0012

Delgado-López, L. (2015). Beyond the Moon agreement: Norms of responsible behavior for private sector activities on the Moon and celestial bodies. In *Beyond the Moon Agreement: Norms of responsible behavior for private sector activities on the Moon and celestial bodies*. Elsevier.

Deplano, R. (2021). The artemis accords: Evolution or revolution in international space law? *International and Comparative Law Quarterly, 70*(3), 799–819. https://doi.org/10.1017/S0020589321000142

Din, A. U. (2022). The Artemis Accords: The end of multilateralism in the management of outer space? In *The Artemis Accords: The end of multilateralism in the management of outer space?* Routledge, Taylor & Francis Group.

Economic, U. S.-C., & Security Review Commission, i. b. (2017). *China's position on a code of conduct in space*. U.S.-China Economic and Security Review Commission.

Gilbert, A. Q. (2023). Implementing safety zones for lunar activities under the Artemis Accords. *Journal of Space Safety Engineering, 10*(1), 103–111. https://doi.org/10.1016/j.jsse.2022.12.007

Gore, A. (1962). International cooperation in the peaceful uses of outer space: Reports of the committee on the peaceful uses of outer space, the world meteorological organisation and the international telecommunication union: UN Doc.AlC.lIPV 1289 (3 December 1962) at 13. https://www.unoosa.org/pdf/garecords/A_C1_PV1289E.pdf

Harris, K. (2022). *Vice President Harris Advances National Security Norms in Space: New U.S. Commitment on Destructive Direct-Ascent Anti-Satellite Missile Testing*. Retrieved from https://www.whitehouse.gov/briefing-room/statements-releases/2022/04/18/fact-sheet-vice-president-harris-advances-national-security-norms-in-space/

Hubbard, B., Margolis, E. A., & Shindell, M. (2024). *Our future in space: Imagining Moon bases, missions to Mars and more.* Raintree.

International Lunar Research Station (ILRS) Guide for Partnership. (2021). China National Space Administration and the State Space Corporationv "Roscosmos" Retrieved from https://www.cnsa.gov.cn/english/n6465652/n6465653/c6812150/content.html

Jakhu, R. S., Nyampong, Y. O. M., & Pelton, J. N. (2016a). Asian space programs: Japan, China and India. In *Space mining and its regulation* (pp. 99–112). Springer International Publishing AG. https://doi.org/10.1007/978-3-319-39246-2_9

Jakhu, R. S., Nyampong, Y. O. M., & Pelton, J. N. (2016b). National space laws and the exploitation of natural resources from space. In *Space mining and its regulation* (pp. 131–144). Springer International Publishing AG. https://doi.org/10.1007/978-3-319-39246-2_11

Jasani, B., & Stockholm International Peace Research, I. (1987). *Space weapons and international security*. Oxford University Press.

Kathryn, L. G. (2022). *China as peer competitor?: Trends in nuclear weapons, space, and information warfare*. Air University Press.

Khalid, M. (2021). Space legal regimes, militarization, and weaponization of outer space. *Astropolitics, 19*(1-2), 128–144. https://doi.org/10.1080/14777622.2021.2008768

Kostenko, I. (2021). Artemis Accords and the future of space governance: Intentions and reality. *Advanced Space Law (Online), 8*, 40–50. https://doi.org/10.29202/asl/8/4

Lee, R. J. (2004). Article II of the outer space treaty: Prohibition of state sovereignty, private property rights, or both? *Australian International Law Journal, 11*(11), 128–142.

Leterre, G. (2024). *Protecting the last frontier: Space mining and environmental sustainability*. Wolters Kluwer.

Mallowan, L., Rapp, L., & Topka, M. (2021). Reinventing treaty compliant "safety zones" in the context of space sustainability. *Journal of Space Safety Engineering, 8*(2), 155–166. https://doi.org/10.1016/j.jsse.2021.05.001

Marboe, I. (2012). *Soft law in outer space: The function of non-binding norms in international space law*. Böhlau Verlag Wien.

McKeown, B., Dempster, A. G., & Saydam, S. (2022). Artemis Accords: Are safety zones practical for long term commercial lunar resource utilisation? *Space Policy, 62*, 101504. https://doi.org/10.1016/j.spacepol.2022.101504

Mercer, A., & Queen Elizabeth, H. (2009). *The role of the global compact in the creation of soft law: the case for soft law in international governance* Thesis (M.Sc. in Global Governance & Diplomacy)--University of Oxford, 2009.].

Meyer, A. T. G. T. L. (2010). International soft law. *Journal of Legal Analysis, 2*, 171–225.

NASA. (1988). *Lunar Helium-3 and Fusion Power*. https://ntrs.nasa.gov/citations/19890005471

NASA. (2019). *ISS Safety Requirements Document*. Retrieved from https://ntrs.nasa.gov/api/citations/20210009936/downloads/SSP%2051721-Baseline.pdf

NASA. (2020a). *The Artemis Accords principles for cooperation in the civil exploration and use of the moon, mars, comets and asteroids for peaceful purposes*. Retrieved from https://www.nasa.gov/specials/artemis-accords/img/Artemis-Accords-signed-13Oct2020.pdf

NASA. (2020b). *Artemis plan: NASA's Lunar Exploration Program overview*. National Aeronautics and Space Administration.

NASA. (2020c, October 27). *NASA, European space agency formalize artemis gateway partnership*. https://www.nasa.gov/news-release/nasa-european-space-agency-formalize-artemis-gateway-partnership/.

NATO. (2019). *NATO's overarching space policy*. Retrieved from https://www.nato.int/cps/en/natohq/official_texts_190862.htm

Pope, R. (2021). Space weapons and the increasing militarisation of outer space: Whether the legal framework is fit-for-purpose. *Auckland University Law Review, 27*(1), 263–301.

Rachel, N. (2021). Artemis accords: A new path forward for space lawmaking? *Adelaide Law Review, 42*(2), 569–580.

Rajagopalan, R. P. (2022). *ASAT weapons: A real threat to the future of Space*. https://www.orfonline.org/expert-speak/asat-weapons.

Riordan, N., Machoň, M., & Csajková, L. (2023). Space diplomacy and the Artemis Accords. *The Hague Journal of Diplomacy, 18*(2-3), 380–408. https://doi.org/10.1163/1871191x-bja10151

Russia Compares Trump's Space Mining Order to Colonialism. (2022). *Moscow times.*. https://www.themoscowtimes.com/2020/04/07/russia-compares-trumps-space-mining-order-to-colonialism-a69901

Salmeri, A. (2023). Introduction. In *The multi-level governance of space mining*. Wolters Kluwer Law International.

Schmitt, H. (2007). *Return to the Moon: Exploration, enterprise, and energy in the human settlement of space*. Springer. https://books.google.co.uk/books?id=IerrQGC6S2YC

Sivolella, D. (2019). *Space mining and manufacturing: Off-world resources and revolutionary engineering techniques* (1st ed.). Springer. https://doi.org/10.1007/978-3-030-30881-0

Space Power*: Doctrine for Space Forces*. (2020).

Space: The $1.8 trillion opportunity for global economic growth. (2024). McKinsey & Company. https://www.mckinsey.com/industries/aerospace-and-defense/our-insights/space-the-1-point-8-trillion-dollar-opportunity-for-global-economic-growth

Takeuchi, Y. (2022). Rights and obligations in international consultation for establishing safety zones in outer space. *International Institute of Space Law, 65*(5), 375–380. https://doi.org/10.5553/IISL/2022065005005

Tronchetti, F. (2008). The non-appropriation principle as a structural norm of international law: A new way of interpreting Article II of the outer space treaty. *Air & Space Law, 33*(3), 277–305. https://doi.org/10.54648/AILA2008021

Tronchetti, F., & Liu, H. (2021). Australia's signing of the Artemis Accords: A positive development or a controversial choice? In *Australia's signing of the Artemis Accords: A positive development or a controversial choice?* Routledge, Taylor & Francis Group.

Trump, D. (2020). *Executive order on encouraging international support for the recovery and use of space resources*. Retrieved from https://trumpwhitehouse.archives.gov/presidential-actions/executive-order-encouraging-international-support-recovery-use-space-resources/

U.S. Commercial Space Launch Competitiveness Act *of 2015: report of the Committee on Commerce, Science and Transportation on S. 1297*. (2015). U.S. Government Publishing Office.

UNIDR. (2007). *Celebrating the space age: 50 years of space technology, 40 years of the outer space treaty: Conference report, 2–3 April 2007. 50 years of space technology*.

Variath, A. (2024a). *Lunar mining and the erosion of the common heritage principle: What's the future of the moon agreement?*. https://cilj.co.uk/2024/10/20/lunar-mining-and-the-erosion-of-the-common-heritage-principle-whats-the-future-of-the-moon-agreement/

Variath, A. (2024b). Space economy and the geopolitics of South Asia's space security trilemma. *Round Table (London), 113*(2), 202–203. https://doi.org/10.1080/00358533.2024.2333201

Variath, A. (2024c). Time for an Indian National Space Security Policy? *Lex ad Coelum, 4*(I).

Vertadier, H., & Kate Schingler, J. (2023). Safety zones for sustainability on the moon: The example of the breaking ground trust. *Air & Space Law*, *48*(Special Issue), 191–210. https://doi.org/10.54648/AILA2023038

von der Dunk, F. G. (2021). The Artemis Accords as a tool of cooperation. *International Institute of Space Law, 64*(2), 145–157. https://doi.org/10.5553/IISL/2021064002006

Wu, X. (2023). The International Lunar Research Station: China's new era of space cooperation and its new role in the space legal order. In *The International Lunar Research Station: China's new era of space cooperation and its new role in the space legal order*. Elsevier.

Xu, F., & Ou, J. (2023). Promoting international cooperation on the International Lunar Research Station: Inspiration from the ITER. In *Promoting international cooperation on the International Lunar Research Station: Inspiration from the ITER*: Published by Elsevier Ltd. on behalf of IAA.

Chapter 8
Deconfliction Measures in the Freedom of Use of Outer Space: What Lies Ahead?

Taufik Rachmat Nugraha and Regi Rivaldi

Abstract The deconfliction principle sets boundaries for the Artemis Accords' parties to ensure the safety of activities and environmental protection on the moon's surface. The essence of the deconfliction principle is to avoid collisions and interference between space activities carried out by its members. The deconfliction principle will raise questions for non-Artemis Accords parties who have and wish to settle activities on the moon's surface in the same area as the deconfliction principle is applied to the Artemis parties since the Artemis Accords is not a legally binding document. Articles I and II of 1967, the Outer Space Treaty (OST), stated that every country has the right to access outer space, including the moon and other celestial bodies, without any discrimination. Furthermore, Art II of the OST states that outer space, including the moon and other celestial bodies, is subject to the non-appropriation principle. However, Article IX of the OST obliged all parties to have due regard principle to avoid harmful contamination effects from space activities. This chapter aims to clarify the relationship between the deconfliction and the general principles outlined in Articles I, II, and IX of the 1967 OST.

Keywords Deconfliction · Safety zones · NASA · Moon Agreement · Space Exploration · Non-appropriation

T. R. Nugraha (✉)
Centre for Air and Space Policy, Surabaya, Indonesia

The Indonesian Centre for the Law of the Sea (ICLOS), Universitas Padjadjaran, Sumedang, Indonesia
e-mail: taufik18004@mail.unpad.ac.id

R. Rivaldi
Universitas Padjadjaran, Bandung, Indonesia
e-mail: regi20002@mail.unpad.ac.id

S. Bhat B., A. Variath (eds.), *Artemis Accords and Resource Mining in Outer Space*, Issues in Space, https://doi.org/10.1007/978-981-95-0152-6_8

8.1 Introduction

Since the dawn of civilisation, outer space has been a frontier of curiosity, exploration, and potential. It has inspired diverse cultures and nations, helping humankind to understand the universe. From fiction to reality, the development of space technology has helped us to explore places where no one has gone before. Since then, outer space has been an object of competition. From the utilisation of geopolitical tools only to increasingly relying on geoeconomics methods for power competition (Dobos, 2022).

The first human race to conquer outer space began with Sputnik I in 1957—the first artificial satellite in outer space launched by the Soviet Union. This achievement immediately raised questions in the international community regarding the definition of the status of outer space and the legal framework governing outer space. The rivalry between two spacefaring nations that were involved in the Cold War—the United States and the Soviet Union—further underscored the necessity for a legal regime to prevent conflicts and ensure the peaceful use of space. The urgency to establish international regulations for the use of outer space led to the development of the Treaty on Principles Governing the Activities of States in the Exploration and Use of Outer Space, including the Moon and Other Celestial Bodies (Outer Space Treaty). However, the Outer Space Treaty was never intended to regulate all aspects of outer space activities. Instead, it was designed to regulate human activities in outer space in general, without providing specific legal definitions or terminology regarding those activities.

Article I of the Outer Space Treaty is a fundamental principle in international space law that provides guidance for States and it has a direct impact on all space activities. It declares that outer space, including the moon and other celestial bodies, is free for exploration and use by all States (United Nations, 1967). This principle ensures that no single nation can monopolise space or its resources. However, the Outer Space Treaty does not explicitly define the terms "exploration" and "use" in Article I. This lack of clarity allows countries to interpret these terms according to their own national interests, including incorporating such interpretations into their national laws.

In April 2020, President Donald Trump signed an executive order to encourage international cooperation in exploring and using resources of outer space. The Executive Order stated that the United States does not view outer space as a part of the global commons and dismissed the relevance of the Moon Agreement as an international treaty and customary international law (Office of Space Commerce, n.d.). Months later, on October 13, 2020, under the administration of President Joe Biden, the United States led the signing of the Artemis Accords with seven other countries to establish a set of principles for peaceful and sustainable cooperation in outer space *exploration* (U.S. Embassy & Consulates in Italy, n.d.). Drafted by the National Aeronautics and Space Administration (NASA), the Artemis Accords are named after NASA's Artemis program, which aims to return humans to the Moon and establish the Moon as a base or outpost to support space exploration missions to the Moon, Mars, and beyond. As of January 2025, 53 countries have signed the Accords and stated their commitment to collaborate in sustainable space exploration (NASA, 2020). However, the Accords have received mixed reactions; some spacefaring nations Russia and China, remaining outside the Accords (Deplano, 2021).

The Artemis Accords represent a significant development in the international landscape of space exploration. Two sections of the Accords have sparked debate among scholars due their implications for international space law, as it directly address space resource extraction: Section 10, which concerns the utilization and property rights of space resources and Section 11, which clarifies the principle of deconfliction of space activities. The Accords are an extension of the Artemis Program, designed to conduct sustainable space exploration activities. The primary purpose of the Accords is to foster international cooperation in lunar exploration activities. This allows nations to pool resources, share their scientific research, and advance lunar science and technology more rapidly than individual efforts alone. The Accords also encourage cooperation and communication to avoid collisions, interference, or other harmful interactions between space activities, particularly in the increasingly crowded lunar environment. This reflects one of the principles in the Artemis Accords known—the deconfliction of outer space activities—as outlined in Section 11 of the Accords. The focus of the principle is to prevent harmful interference and potential conflict among the States Parties.

The deconfliction principle is not explicitly stated in the Outer Space Treaty. This principle encourages States to plan their space activities to avoid collision, interference or other harmful interactions. This collaborative approach aims to prevent harmful interference with the activities of other States Parties or collisions that could jeopardize the success and safety of space missions. To implement this principle, Section 11, Paragraph 5 of the Artemis Accords requires a commitment from States Parties to exchange and provide information about the location and their space activities to establish the safety zones and prevent potential conflicts to avoid harmful interference (NASA, 2020, Section 11 Paragraph 5). The concept of safety zones is mentioned in Section 11 Paragraph 2 of the Artemis Accords and it does not represent an entirely new concept in the field of international law. The concept has been further explored and discussed by the Hague International Space Resources Governance Working Group with the aim of ensuring the long-term sustainability of outer space activities and it aligns with the UN Long-term Sustainability of Outer Space Activities. The concept of safety zones in the Artemis Accords has two main purposes: to protect ongoing scientific research from interference and to support the extraction of space resources (Deplano, 2021). Notably, the concept of safety zones does not appear explicitly in the Outer Space Treaty.

The Outer Space Treaty is the foundation of international space law, and it directly shapes all aspect in space activities. However, it is important to note that the Outer Space Treaty is a set of broad frameworks for space activities, and it was written in general terms without providing specific regulations, which leaving room for interpretation. The Treaty governs human activities in outer space, including on the Moon and other celestial bodies. Articles I, II, VI, VII and IX establish a framework regulating the exploration and use outer space, ensuring that activities are conducted for the benefit of all mankind. Article I of emphasizes the freedom of exploration and use of outer space by all nations, while Article II prohibits national appropriation of outer space, ensuring that the Moon and other celestial bodies cannot be claimed by any country. Article VI holds States responsible for national activities in space, including those conducted by private entities, reinforcing the need for oversight and accountability. Article VII addresses liability for damage caused by man-made space objects, ensuring that States bear responsibility for their space activities. Finally, Article IX focuses on preventing harmful contamination and requires States to avoid harmful interference with the space activities of other countries.

This chapter attempts to clarify the relationship between the deconfliction principle and the fundamental principles in the Outer Space Treaty and how far the deconfliction principle affects the freedom of use of outer space.

8.2 Deconfliction Principle Under International Space Law

The Artemis Accords represent a significant shift in the utilisation of outer space, particularly in the exploration and exploitation of space resources. However, it is important to note that the intent of the Artemis Accords extends beyond resource exploitation. A key addition, outlined in Section 11 of the Artemis Accords is the concept of 'Deconfliction of Space Activities' also known as 'Safety Zone.' This concept whether referred to as deconfliction or a safety zone is a novel inclusion in space law regimes under the Outer Space Treaty. It is also worth noting that the concept of safety zones is referenced in the 1979 Moon Agreement; however, the agreement has limited acceptance, with only 17 signatory countries, and is generally considered ineffective.

The Artemis Accords underscore the pivotal role of the deconfliction zones in ensuring safety and preventing harm to all parties engaged in activities on the Moon and other celestial bodies. This is achieved through timely notification and implementation of the deconfliction measures. NASA has emphasised that these measures will provide comprehensive information about ongoing activities in the surrounding areas, thereby establishing safety zones.

However, the deconfliction principle originates from international humanitarian law, where it was developed as an Ad Hoc military security procedure aimed at preventing accidental attacks and killing of soldiers belonging to the same or allied parties in an armed conflict.

Furthermore, the core of the deconfliction principle lies in the exchanging information to prevent hazardous outcomes resulting from activities in specific areas. In addition, the deconfliction principle is also known to make a 'safety zone' that has never existed before in space law. Safety zones first appeared in the context of space law through Building Blocks for Development of an International Framework on Space Resources Activities 2019. This document emphasizes that safety zones are intended solely to prevent hazardous impacts during space activities particularly in relation to In-Situ Resource Utilization (ISRU) and are not meant to impede free access to outer space by other actors.

The foundation of safety zone in the Artemis Accords Section 11, Paragraph 3 aims to align with Article IX of the Outer Space Treaty: "…If a State Party to the Treaty has reason to believe that an activity or experiment planned by it or its nationals in outer space, including the Moon and other celestial bodies, would cause potentially harmful interference with activities of other States Parties in the peaceful exploration and use of outer space, including the Moon and other celestial bodies, it shall undertake appropriate international consultations before proceeding with any such activity or experiment…."

Art IX of the Outer Space Treaty, as reflected in the Artemis Accords emphasises if an activity is believed to cause harmful interference and requires the establishment of a safety zone, this should be determined through international consultations, rather than through unilateral action. However, the term 'international consultation' raises questions regarding the specific procedures and the entities involved in such a process. In his book, Bin Cheng notes that, to implement Article IX of the Outer Space Treaty, a State Party must have "reason to believe" that its activity may cause harmful effects. He further argues that international consultation is not a strict prerequisite before a party proceeds with an activity. This suggests that the international consultation process may not be clearly defined, leaving room for interpretation and potential disputes.

While the deconfliction principle is intended to ensure the safety of operations within specific areas, it may also raise concern regarding the freedom of access to outer space. This concern aligns with Article I of the Outer Space Treaty, which states that outer space, including the Moon and other celestial bodies, shall be free for exploration and use by all States without discrimination of any kind. Additionally, it will intersect with the non-appropriation principle, as Article II of the Outer Space Treaty states that the moon and other celestial bodies are not subject to national appropriation.

The issue of access freedom in outer space, including the Moon and other celestial bodies, remains a central concern. Since Article I of the Outer Space Treaty guarantees this freedom to all nations, ensuring that space shall be free for exploration and use without discrimination. However, the implementation of the deconfliction principle—particularly through the establishment of safety zones—may raise questions about its potential impact on this guaranteed freedom. This stands in contrast to the situation with the Geosynchronous Orbit (GSO), which, although limited in capacity, is not subject to appropriation by states that place satellites there. In this context, safety zones could be seen as functionally restrictive, even if not formally violating the non-appropriation or access principles of the Outer Space Treaty.

However, the deconfliction principle needs to be addressed in Article IX. In that case, every country shall exchange the information with all countries to prevent any hazardous contamination over some area based on the due regard principle. If further worries arise, the deconfliction principle will blanket state parties to Artemis Accords to "own" some areas regarding the safety of operation measures. However, in theory, such an act will breach Article I of the Outer Space Treaty, that outer space shall be free access for all countries, and in Article II of the Outer Space Treaty, space object is not subject to appropriation since the practice may be contrary to the theory.

However, Article II of the Outer Space Treaty does not specify the time. Meanwhile, the deconfliction may be applied temporarily. Refer to the Russian proposal on establishing the Safety Zone, a "specific form of temporary sovereign rights in a limited area of near-earth space." However, regarding sovereign rights on the Law of the Sea, passage will be permitted under specific terms. Nevertheless, under Art I 1967 OST, such sovereign right is prohibited by all means since outer space status is under the province of all [hu]mankind.

Whether the Artemis Accord fully comply with the Outer Space Treaty remains a subject of debate—particularly regarding Section 11, Paragraph (d), which states that the implementation of safety zones should be "...consistent with Article XI of the Outer Space Treaty...." In practice, this provision may lead to confusion or concern, as it risks creating a form of "pseudo-sovereignty." to situations in which State Parties establish deconfliction zones unilaterally—based solely on their own studies and without third-party validation—effectively preventing other actors from accessing the same area or areas within the zone's perimeter. As for comparison, in the Area, a safety zone can be established for the safety of navigation, yet it should not impede access to the freedom of navigation.

Argumentum per Analogiam under the Area regimes is established in Chapter 11 of the LOSC and the Implementation Agreement 1994, designating the Area as the Common Heritage of Mankind (CHM), as articulated in Article 136 and Article 137. Within this regime, no state may assert sovereignty or sovereign rights over any part of the Area. However, Chapter XI provides a legal framework for the exploration and exploitation of resources in the Area, carried out under the supervision and authority of the International Seabed Authority (ISA) Furthermore, the implementation of exploration and exploitation in the specific zone where the ISA has jurisdiction does not establish any form of "Pseudo-Sovereignty" over the safety zone as long as all parties adhere to the regulations set forth by the 1982 LOSC and the ISA. Additionally, an international organisation is in place to prevent excessive claims by state parties over specific areas. In the context of the Artemis Accords, it is noted that compliance is essential, and following the Outer Space Treaty, there is no particular authority to ensure the Accords meet their obligations under international law. Conflicts may arise without regulatory bodies similar to the ISA in the Area regimes.

Moreover, the concept of "Pseudo-Sovereignty" can be illustrated through the Arctic Sunrise case. This involved a Greenpeace vessel protesting against Gazprom's drilling operations at the Prirazlomnaya oil platform. The protest occurred just three nautical miles from the platform, within the Exclusive Economic Zone of the Russian Federation. s the EEZ is considered a hybrid legal zone, the high seas regime continues to apply within it—allowing for certain freedoms of navigation and activity—except where such freedoms are limited by the rights of the coastal state under Article 56 of the 1982 LOSC. These rights include, among others, the construction of artificial islands, marine scientific research, and the protection and preservation of the marine environment. In the context of the Arctic Sunrise case, it is evident that the protester did not violate Article 58 of the 1982 LOSC. Furthermore, the conduct did not contravene Article 88, which reserves the high seas for peaceful purposes, nor Article 111, which governs the right of hot pursuit.

Refer to Section 11 Paragraph 7(d), the state parties shall inform the United Nations Secretary-General of the establishment, alteration or end of any safety zone in line with Art XI 1967 OST. However, the primary function of Art XI is to disclose information regarding the activities or the result of such activities. Nonetheless, the wording in Art XI, "…to the greatest extent feasible and practicable…" has much limbo on which information the state parties could disclose the information.

The second key question regarding the deconfliction measure is whether this measure will apply exclusively to the parties of the Artemis Accords, or whether it will be imposed on all nations, including those that are non-parties to the Accords. Given the Artemis Accords are politically binding rather than legally binding under international law, it becomes problematic to expect non-party states to adhere to obligations that they have not formally accepted. This raises fundamental questions about the legitimacy and authority of applying such measures universally. The fundamental argument of the deconfliction principle is how to use this principle appropriately during its practice.

According to Section 7, there are some points regarding implementing a deconfliction zone, such as the determined area shall be reasonable, the perimeter is expected to change due to the operational zones, and the safety zones are ultimately temporary.

8.3 Pseudo-Sovereignty in Space Exploration and Use

The concept of pseudo-sovereignty could become an important issue as space exploration continues to advance. Pseudo-sovereignty refers to a deceptive condition in which one party unilaterally controls a specific area in space under *res communis* regimes by establishing boundaries that other entities are not allowed to cross. This is typically done based on the controlling party's safety concerns and beliefs, without proper consultation with the international community.

According to Miriam Webster, the word "pseudo" originates from the Greek term *Ψευδής*, meaning false. Essentially, "pseudo" denotes something that is assumed rather than actual (Merriam-Webster, n.d.). Sovereignty, as per Bodin, refers to the absolute and perpetual power of a commonwealth (Bodin, 1992) and the totality of international rights and duties recognised by international law as residing in an independent territorial unit (Crawford, 1994, p. 1). In 1967, OST, humans have much freedom to explore and use outer space. However, freedom comes with limitations. Art I of the 1967 OST stated that access should be freedom for any state, and Art II notions that space is not owned by anyone but by everybody. This notion wants to enlighten that space shall own and be governed under a more significant community than a state, or in short, should be governed and owned by humankind. By declaring a deconfliction zone or safety zone on the moon's surface or other areas that have surfaced in outer space without any international authority particularly monitoring and supervising the activities, it would be hard to secure 'freedom' of access.

Since jurisdiction is part of sovereignty, its shall be clear where the jurisdiction is placed under space law objects, yet there are insufficient and neither clear nor satisfactory definitions of space objects, yet we could refer to the 1975 Registration Convention, Art I (b) includes the component parts of a space object as well as its launch vehicle and parts thereof. Furthermore, Art IV 1967 OST stated that "…state parties…undertake not to place in orbit around the Earth any objects…." The definition shall be clear since the flag state has quasi-jurisdiction over its space objects. Moreover, there will be a question of whether the jurisdiction is solely limited to the space objects or whether it could be extended into the area surrounding it if there are installations for mining.

For example, under Art 91, LOSC states, "…every state shall fix the conditions for the grant of its nationality to ships for the registration of ships in its territory…." Furthermore, in Art 92 of LOSC, "…shall be subject to its exclusive jurisdiction on the high seas…." The nationality of ships is important because the exclusive flag state jurisdiction allows ships to be controlled in their activities (Guilfoyle, 2017, p. 700; O'Connell, 1984, p. 800). In this manner, when the ship is sailing on the high seas, the authority shall be granted to its flag state, and no other flag states could board the ship unless there are specific circumstances, e.g. ship under hot pursuit. Flag states have full control over that ship.

In 2020 China's lunar mission *Chang'e*-5 did flag planting on the lunar surface, yet sometimes this thing is often dismissed by the international community as a 'regular event,' we have to carefully understand that China has a 1990 Flag Law that deface or tarnishes of Chinese flag is illegal act (Lim, 2020). According to Jon Lim's analyses, the Chinese flag is strongly intertwined with national security matters, and it possibly can extend its jurisdiction to anyone outside the region, non-citizens from the region may face prosecution, or there is no place hiding in this world from such a law. Furthermore, since the extended jurisdiction is engaged with its jurisdiction, it's also possible to claim a 'safety zone' by planting their flags in the Lunar region and making that area "belong to China" and if this practice is broadly accepted by the international community, it will causing "pseudo-sovereignty" over the lunar surface (Lim, 2020).

8.4 The Enigma of Space Exploration and Use

8.4.1 The Terms of the Freedom of Exploration

The potential for commercial use of outer space has sparked a debate over the extent to which public and private property rights can be acquired over space resources under current international space law. At present, there is no definitive definition or interpretation on this matter. It remains unclear whether the acquisition of property rights over the natural resources of the Moon and other celestial bodies is permissible. The Outer Space Treaty, by addressing only general aspects of space activities, is limited in its capacity to provide detailed guidance on how specific space-related activities should be conducted.

Article I of the Outer Space Treaty is often described as the most fundamental principle in international space law, directly influencing all related space activities through the principle of freedom of exploration and use of outer space (Tronchetti, 2009, p. 20). The provisions in Article I of the Treaty grant all countries the right to engage in outer space exploration, including on the Moon and other celestial bodies. Article I (2) defines three basic rights: the right to free access, the right to free exploration, and the right to free use (Tronchetti, 2009, p. 22). These rights ensure that all nations have equal opportunities to access and utilise outer space, regardless of their economic capabilities. This principle, often referred to as the "province of all mankind," asserts that the exploration and use of outer space should benefit all countries and humanity as a whole. However, the terms "exploration" and "use" are not explicitly defined in detail within the provisions of Article I of the Outer Space Treaty.

Article I of the Outer Space Treaty does not clarify whether the terms "exploration" and "use" encompass scientific research and the extraction of space resources. Additionally, the text does not address the acquisition of property rights over extracted resources, which falls under the non-appropriation principle outlined in Article II of the Treaty. Furthermore, numerous questions currently under discussion concern the exploration, use, and legal status of outer space, as well as the implications of these activities, and the extent to which the principles of freedom of exploration and use regulate human activities in outer space.

An international agreement functions as a tool to align the interests of different states on a specific issue. While the signing and ratification of such treaties are not always mandatory, these actions represent a formal process to ensure state compliance with the treaty's provisions. Despite the clear language and terminology used in these treaties, many states tend to interpret them in ways that serve their national interests. The Outer Space Treaty serves as a foundation of international space law, governing human activities in outer space. Although the principles of freedom of exploration and use, as well as the non-appropriation principle, have been widely embraced by the member states of Outer Space Treaty and have evolved into customary international law, their practical application reveals significant variations in interpretation.

In order to understand the significance of the treaty, it is essential to interpret its provisions clearly to obtain a precise understanding of its meaning. Interpretation is a crucial element in understanding and implementing treaties, ensuring that the intentions of the involved parties are upheld and that the agreements operate as intended, as outlined in the Vienna Convention on the Law of Treaties (VCLT), specifically in Article 31 (Aust, 2013, p. 218). Article 31(1) of the VCLT states that a treaty must be interpreted in good faith, a provision derived from the principle of *pacta sunt servanda*, as set forth in Article 26 of the VCLT 1969 (United Nations, 1969, Art. 31). Shifts in meaning when interpreting the content of an agreement from its original meaning are common when interpreting the provisions of an international agreement. The Vienna Convention on the Law of Treaties of 1969 regulates how the provisions of a treaty can be interpreted. Widely accepted by the international community, this Convention has become part of customary international law regarding treaty interpretation.

Article 31 of the VCLT emphasises that the terms of a treaty should be interpreted according to their ordinary meaning, understanding the words as they are commonly used unless the treaty explicitly defines them otherwise. However, there is no universal consensus on the specific meanings of "the freedom of exploration," "use," or "scientific research" outlined in Article I of the Outer Space Treaty. This approach offers limited insight into the ordinary meaning of the term "use" in Article I of the Outer Space Treaty. Specifically, it does not clarify whether the term "use" includes the activities of the extraction of space resources or extends to commercialisation activities.

Article I of the Outer Space Treaty implies that space activities should be conducted for the benefit of all countries and should be of concern to all humanity. However, Article I does not specify whether property rights can be asserted over extracted space resources. Some experts argue that the stipulations of Article I can be interpreted as "legally binding behavioural imperatives" intended to guide, influence, and shape the nature of space activities in a specific manner (Tapio, 2024, p. 68). The article ensures that the benefits of space activities should be determined through international cooperation, allowing equitable sharing among all nations, regardless of their economic and technological development. Space is viewed as a *global commons* that should be shared and governed collectively by the international community.

Historically, the development of international space law is deeply entwined with the political context in which the Outer Space Treaty was conceived and negotiated. The Treaty was crafted during a period of intense geopolitical tension—commonly referred to as the Cold War. This era, characterised by the space race between the United States and the Soviet Union, compelled the international community to establish legal frameworks to prevent the militarisation of outer space and to ensure its peaceful use. The Outer Space Treaty emerged in response to the urgent need of the international community for regulations governing human activities in space. However, due to the relatively slow pace of space technology development at that time, the exploitation and utilisation of space resources were not a primary focus for the drafters. Understanding the Outer Space Treaty requires consideration of the technological advancements and political dynamics of its time. As the political landscape evolves and space technology advances rapidly, there has been a shift in the interpretation of the provisions of the Treaty by nations. One example is the principle of non-appropriation, which was originally interpreted broadly but has since narrowed in scope, diverging from the original intent of the drafters.

Article I (2) of the Outer Space Treaty confirms that outer space falls under the *res communis omnium* regime (Tronchetti, 2009, p. 21). The origin of this concept lies in Roman law, specifically in the writings of the jurist Marcianus, preserved in the Roman legal texts known as the "*Digest*" or "*Pandects*" (*Digesta seu Pandectae*).

These writings contributed significantly to the preservation and codification of Roman legal principles and remain an important source in the history of Roman law. These texts refer to the use of the open sea as a resource available to all without exclusive rights for any individual (Fenn, 1925). In the seventeenth century, the concept of *res communis omnium was further developed in Hugo Grotius' book "Mare Liberum", which asserted that the open sea is an area open to all countries. This principle was later codified into the international legal system in Article 87 of the 1982 Law of the Sea Convention (LOSC), granting states the right to fish on the high seas, and in Article 137 of the LOSC, which prohibits any state from asserting or exercising sovereignty or sovereign rights over any portion of the deep seabed and its resources.*

Additionally, this concept was incorporated into international space law through Article II of the Outer Space Treaty, which stipulates that outer space is governed by the principle of non-appropriation. However, due to their unique characteristics, areas designated as *res communis omnium*—such as Antarctica, the high seas, and the seabed – are primarily governed by public international law. This raises questions about the legal status of natural resources within the *res communis omnium* areas, including the utilisation of natural resources in outer space. In the historical context of space law, the concept of *res communis omnium* garnered significant support, largely due to the geopolitical conditions of the time.

The principle of freedom of exploration and use, as articulated in Article I of the Outer Space Treaty, imposes legally binding obligations on States Parties under international law. It is widely believed that the principle has evolved into customary international law since its adoption in the United Nations General Assembly's resolution 2222 (XXI) (UNGA Res 2222) on 19 December 1966. The primary purpose of the provisions in Article I is to establish a legal regime for outer space that regulates not only the rights of states directly involved in space activities, but also the rights of all mankind (Tronchetti, 2009, p. 21). The convention has been developed and implemented through three different conferences, often referred to as UNCLOS I, II, and III; the last one being the most comprehensive and widely recognised, which established the current Law of the Sea Convention.

Unlike the LOSC, which provides specific definitions for the terms "exploration" and "exploitation," the Outer Space Treaty does not offer detailed definitions for the terms "exploration" and "use," leaving room for varying interpretations. The LOSC defines the term "exploration" with a focus on regulating commercial activities and the exploitation of natural resources, whereas the Outer Space Treaty places greater emphasis on scientific aspects. This distinction reflects the different objectives of the two legal instruments: UNCLOS III (1982) aims to regulate the fair and sustainable use of natural resources, whereas the Outer Space Treaty aims to ensure that space exploration and utilisation are conducted for peaceful purposes and for the benefit of all humankind. Consequently, interpreting the term "use" through the method of ordinary meaning presents significant challenges.

Article I of the Outer Space Treaty establishes a fundamental principle that influences all state activities in outer space—a principle widely recognised by states and international legal experts as part of customary international law. For a practice to attain as a customary international law, it must satisfy two key elements: general practice, meaning the repeated actions of the majority of states, and *opinio juris*, the belief that such practice is legally obligatory. The United States has sought to establish customary international law in utilisation over space resources through the Artemis Accords, which seek to create it rapidly and implicitly—a concept Bin Cheng refers to as "instant customary international law" (Cheng, 1965). Bin Cheng proposed this theory in response to UN General Assembly Resolutions No. 1721 and No. 1962 (UNGA 1721), which concern the principles of law governing space activities. According to Cheng, the practices outlined in these two resolutions had not yet met the requirements, such as the duration and the breadth of practice, to be considered customary international law (Cheng, 1965). Nonetheless, the practices of the United States and the Soviet Union, along with the principles enshrined in these resolutions, were accepted by other countries in a relatively short time. Therefore, Bin Cheng argues that as long as the *opinio juris* regarding a norm in international law is not rejected by states, customary international law can be formed instantly (Cheng, 1965).

The Vienna Convention specifies that, when interpreting treaties, the scope includes not only the main text but also the preamble and annexes. Therefore, when interpreting Articles I and II of the Outer Space Treaty, it is essential to consider other relevant provisions within the treaty itself. Article VI of the Treaty addresses the involvement of non-state actors in space activities. According to this article, activities conducted by non-governmental entities in outer space must be authorised and continuously supervised by the appropriate State Party to the Treaty. Moreover, a state's jurisdiction extends beyond its territory and nationals to include activities involving objects registered within its borders. Consequently, state responsibility in space encompasses not only the space activities carried out by the state itself but also those conducted by non-state actors.

Article I of the Outer Space Treaty grants no exclusivity limited to states having the capability to explore outer space, as well as protecting the interests of non-spacefaring nations. Additionally, outer space is widely recognised as *res communis omnium.*

8.4.2 Interpreting the Provisions of the Outer Space Treaty

The discussion surrounding the utilisation of outer space for commercial ventures has ignited a debate on the feasibility of acquiring public and private property rights over space resources within the framework of existing international space law. Various interpretations exist, but there is no universally accepted definition or consensus on has yet emerged. In particular, the permissibility of acquiring property rights over the natural resources located on celestial bodies, such as—the Moon—remains ambiguous under current legal regime.

Interpretation is a fundamental method for understanding the meaning of an international treaty and for creating a clear and coherent system to comprehend its contents (Aust, 2013). Traditionally, the process of interpreting international law has focused on understanding the text and its implications for legal rights and obligations (Herdegen, 2013, p. 123; Bianchi et al., 2015, p. 3). Furthermore, the Vienna Convention allows "supplementary means of interpretation" including the preparatory work of the treaty and the conditions of its signing, which can be used to confirm or clarify the meaning of the treaty only if the text of the treaty is unclear or leads to an unreasonable outcome (Byers & Boley, 2023, p. 25).

The successful Moon landing of the Apollo 11 mission marked a significant turning point in how spacefaring nations interpreted the non-appropriation principle. The collection of lunar samples taken from the Moon by Neil Armstrong was widely perceived as an implicit assertion of ownership over the lunar material brought back to Earth by the United States. The Soviet Union undertook similar actions. Since that time, ownership claims of lunar samples by the United States and the Soviet Union (and later Russia) have not been seriously contested, suggesting an indirect acceptance by other states of property rights over extracted materials from celestial bodies. Following the Apollo 11 mission, the United States government collected 842 pounds of lunar material, which, together with other space materials collected by American astronauts, is explicitly regarded as government property (NASA Office of Inspector General, 2011, p. 8). This practice was adopted by the Soviet Union/Russia, indicating a shift in the interpretation of the non-appropriation principle in Article II of the Outer Space Treaty by various countries. Furthermore, the lack of challenges to the ownership claims of lunar samples by the United States and the Soviet Union/Russia indicates tacit recognition by the international community of the right to possess such extraterrestrial materials.

To legitimise its interpretation of Article II of the Outer Space Treaty, the United States seeks to have its understanding of the article recognised as "subsequent practice." One avenue for this is through the Artemis Accords. Section 10, Paragraph 2 of the Artemis Accords states that the extraction of space resources does not inherently constitute national appropriation under Article II of the Outer Space Treaty. In simpler terms, the Artemis Accords argue that extracting resources from space does not violate the non-appropriation principle. However, this interpretation is controversial, with some legal scholars asserting that any resource extraction could be considered a form of *de facto* appropriation.

The interpretation of the non-appropriation principle in Article II of the Outer Space Treaty by the United States and other signatories of the Artemis Accords raises a critical question: Can the signing of the Artemis Accords by 33 members of the Outer Space Treaty be considered as "subsequent practice" that alters the interpretation of the non-appropriation principle and allows private entities and individuals to claim ownership? Given that the concept of "national appropriation" does not have an ordinary meaning and is not commonly used in other areas of international law or everyday conversation, interpreting this term through the ordinary meaning method is difficult (Byers & Boley, 2023, p. 140).

Article 31 of the VCLT provides guidance for examining a treaty within the context in which it was negotiated. This includes considering the treaty's preamble, other articles, annexes, and any relevant agreements or understandings reached during the negotiation process. In the historical context, the Outer Space Treaty was signed during the Cold War era—a period of intense competition between the United States and the Soviet Union. At the time of the negotiation of the Treaty, the principle of non-appropriation was relatively uncontroversial, as the two dominant spacefaring nations—the Soviet Union and the United States—had already recognised this concept as a fundamental principle. This principle emerged to safeguard outer space, the Moon, and other celestial bodies as areas or territories that could not be claimed by any nation, thereby preventing military action. Furthermore, the limited technology and the small number of actors capable of engaging in space activities at the time led to the drafting of the treaty under the assumption that the United States would be the sole actor in outer space (Pop, 2000, p. 276).

The United States explicitly asserts that the principle of non-appropriation applies only to natural resources located in their original place (*in situ*). However, it also argues that once natural resources have been extracted (taken) from their original place, they are no longer bound by the principle of non-appropriation and can be traded commercially.

The United States' interpretation of the non-appropriation principle is clearly reflected in its domestic legislation and policies that support this stance, including the Artemis Accords. 53 countries have expressed their willingness to join the Artemis Accords and participate in the Artemis Program. Notably, the majority of these Accords members are also parties to the Outer Space Treaty. The actions of the Accords' members raise the question of whether the signing of the Artemis Accords by the Outer Space Treaty members can be considered a subsequent practice that could alter the interpretation of the term "non-appropriation" as contained in Article II of the Outer Space Treaty.

In 1978, US Secretary of State Cyrus Vance informed the Senate Foreign Relations Committee that the principle of non-appropriation applies only to natural resources in space if they remain undisturbed in their original location (Israel, 2019). Furthermore, the Secretary stated that this principle does not restrict the property rights of states or private entities over natural resources that have been extracted from their original location or are located beneath the surface of the Moon or other celestial bodies (Pop, 2000, p. 276).

Subsequent practice is a method used to interpret international agreements. In international space law, subsequent practice has played a crucial role in shaping the Outer Space Treaty. The practices or actions of states, whether directly or indirectly involved in space exploration, supported the principle of non-appropriation in outer space, which eventually became a fundamental principle enshrined in the Treaty. The International Law Commission (ILC) considers subsequent practice as a legitimate means to interpret treaties, as outlined in Article 31(3)(b) of the VCLT. This article emphasises the importance of understanding whether the parties to a treaty have reached an agreement or adopted specific positions in interpreting its provisions (ILC, 2018, p. 23).

In the context of the Artemis Accords, no provision explicitly states that signatory countries agree that the interpretation of the provisions in Section 10, Paragraph 2, is the most appropriate interpretation of Article II of the Outer Space Treaty.

Instead, Section 10, Paragraph 2, of the Artemis Accords includes a phrase stating that any legal documents or other agreements governing the utilisation of space resources must be consistent with the Outer Space Treaty. This creates ambiguity in interpreting the provisions of Section 10, Paragraph 2, of the Artemis Accords. Rossano Deplano argues that Section 10 of the Artemis Accords, as a form of subsequent practice, does not accurately reflect the interpretation of Article II of the Outer Space Treaty (Deplano, 2021, p. 9). However, the silence of one or more parties to the Outer Space Treaty could be interpreted as acceptance of the subsequent practice introduced by the implementation of the Artemis Accords, potentially contributing to the formation of *opinio juris*, which is relevant to establishing a customary rule (Deplano, 2021, p. 9). However, given the negative reactions from China and Russia, this scenario seems unlikely at present (Stirn, 2020). Nevertheless, it is worth noting that Russia, alongside China, is collaborating on the International Lunar Research Station (ILRS) project, similar to NASA's Artemis Program, which aims to establish a lunar base near the Moon's South Pole. The ILRS was formed to utilise space resources, particularly those at the Moon's South Pole, emphasising the importance of international cooperation and collaboration in space exploration activities.

8.5 The Rise of Non-state Actors in Space Mining: A New Frontier

The 2015 US Commercial Space Launch Competitiveness Act or "Spurring Private Aerospace Competitiveness and Entrepreneurship Act of 2015" (2015 Space Act), recognises the right of individual and private entities to acquire ownership and commercially exploit space resources, provided these activities comply with US law and do not violate international law. The 2015 Space Act was followed by an Executive Order on "Encouraging International Support for the Recovery and Use of Space Resources" that stated that the United States does not view outer space as part of global commons and dismissed the relevance of the Moon Agreement as an international treaty and customary international law. Furthermore, both regulations are paving the way for the commercialization of space activities, opening doors for private companies to potentially mine asteroids and other celestial bodies for valuable resources. While private space ventures hold substantial potential, several challenges remain. The issue of space resource rights requires further international cooperation to establish a clear legal framework. One-sided claims by private entities could lead to disputes and potentially unsustainable resource extraction practices. Nevertheless, countries such as Japan, Luxembourg, and the United Arab Emirates are taking a proactive stance by establishing legal frameworks to govern potential space mining activities.

International space law does not mention explicitly the term "commercialisation," nor provide specific details regarding space commercialisation. However, there is no provision in international space law that prohibits commercialisation activities in outer space. Space commercialisation refers to activities related to the provision of products or services in outer space by private entities (Vedda, 2004). It holds the potential to democratise access to space and accelerate innovation. Private space companies have advanced the commercialisation of space technology by developing and selling launch services, satellite technologies, and space exploration tools.

Commercialisation is closely related to privatisation, although refer to different aspects of economic and organisational processes. Privatisation, with its many definitions, can occur in various forms. It is often discussed in terms of economics or public policy, and has significant legal impacts (Mills, 2023). Mary M. Shirley's definition of privatisation focuses on the transfer of ownership of assets from the public (government) sector to the private sector (Shirley, 1992). Ownership rights or property rights grant three main rights to the holder: the right to exclusive ownership, the right to use their property, and the right to transfer ownership to another person of their choosing, including determining when the ownership rights are granted (Baca, 1993). These rights play a significant role in the economic system as they empower individuals and organisations to adapt their ownership in response to changing needs and circumstances. Furthermore, they facilitate the efficient exchange of ownership, contributing to enhanced productivity and overall economic well-being. In addition, privatisation and commercialisation are correlated—they are motivated by profit and competition, leading to innovations and market expansion.

Although there is no single provision in the international space law that governs specifically commercialisation of space activities, Article VI and Article VII of the Outer Space Treaty set out the responsibilities and liabilities of states for national activities in outer space, including those carried out by private companies. These provisions establish the framework for how states should oversee and regulate space activities conducted by governmental or non-governmental entities. When a private company engages in space activities, the launching state is responsible for ensuring compliance with the treaty and other relevant international laws. However, the Treaty does not address the accountability of private entities when outer space activities result in harm to third parties. The provisions in these two articles play a crucial role in shaping the legal regime governing the exploitation of outer space resources.

International space law is primarily established through treaties and conventions, and it has traditionally focused on regulating the activities of nation-states in outer space. However, the evolving landscape of space exploration has brought non-state actors into prominence. Non-state actors, in the context of international space law, refer to entities other than sovereign states that engage in activities related to outer space. These can include commercial companies, international organisations, non-governmental organisations (NGOs), and even individuals. Unlike states, non-state actors do not possess the legal personality or sovereignty recognised under traditional international law. At this time, non-state actors' activities in space are governed by national regulations. When the treaty was being negotiated, the possibility of mining space resources and the role of non-state actors were not actively considered by the drafters.

Sovereignty and property rights fall under different branches of the law (Pop, 2000). Sovereignty is a concept rooted in public international law, while property rights originate from private civil law, which governs relationships between individuals (Barnes, 2009, p. 11). In international law, sovereignty is based on the principle that each state is equal and has authority over everything within its territory (Malanczuk, 1997, p. 17). The term "sovereignty" was first coined by Jean Bodin, who defined it as the absolute and perpetual power of a state. According to Bodin, sovereignty is characterised by its indivisibility and supreme authority over all matters within a state (Beaulac, 2003). On the other hand, property rights are part of private law, governing the relationships between legal persons regarding certain tangible and intangible objects (Barnes, 2009, p. 11). Property rights confer three powers to the rights holder: the power to exclude others from the property (exclusive possession), the power to use the property, and the power to transfer the property to another of their choosing at the time of their choosing (Epstein, 1990). For property rights to exist, they require a superior authority to enforce them, whether that be a state or another recognised entity. Without such recognition and enforcement, property rights can only be considered as physical possession without legitimacy. Since Article II of the Outer Space Treaty prohibits sovereignty in outer space, such recognition and enforcement are not feasible. In other words, property rights or claims over land in outer space cannot legally exist within the current framework of international law.

The Moon Agreement, formally known as the "Agreement Governing the Activities of States on the Moon and Other Celestial Bodies," is one of five international treaties that make up the body of space law (United Nations, 1979). The Moon Agreement aimed to provide clearer guidelines on the rights and obligations of states regarding the extraction and ownership of space resources (McKeown et al., 2022). However, since this agreement was opened for signature, none of the space-faring nations have participated in it. One reason for the reluctance of space-faring countries to join this agreement is the provision in Article 11(2), which contains the principle of the common heritage of mankind. Despite the fact that these regulations cover all aspects related to the use of natural resources in outer space, the low interest of countries in ratifying the agreement creates legal uncertainty regarding the exploitation and utilisation of natural resources on the Moon and other celestial bodies. This legal uncertainty is closely related to the evolving economic and political contexts that extend beyond the original scope of the common heritage of mankind principle, including the evolving interpretation of the non-appropriation principle in Article II of the Outer Space Treaty.

The provisions of Article II of the Outer Space Treaty have been interpreted in various ways by legal scholars. Some argue that Article II does not explicitly prohibit private appropriation of outer space; it only forbids national governments from appropriating outer space and does not specifically forbid other entities from doing so (Gorove, 1968, p. 340). When the Treaty was negotiated, the significant role that non-state actors would later play in space activities was not foreseen. However, the absence of explicit references applying the principle of non-appropriation to private entities does not mean that the provisions in Article II of the Outer Space Treaty permit private operators to acquire ownership and property rights over outer space or its natural resources.

As discussed above, Article II of the Outer Space Treaty clearly prohibits any extension of national sovereignty in outer space, including the Moon and other celestial bodies. While all states agree that they cannot extend their jurisdiction to outer space, the debate over the extent to which this principle applies to the extraction of natural resources in outer space remains unresolved. This legal vacuum has been exploited by states to interpret the non-appropriation principle in ways that align with their own interests, including those states that have joined the Artemis Program.

Section 11(5) of the Artemis Accords states that participating members should establish safety zones around their activities to avoid potentially harmful interference. Although these zones are temporary and intended to provide protection for a specific period, they might necessitate a level of control and regulation that could be interpreted as national jurisdiction or appropriation, potentially conflicting with the requirement for non-appropriation. The Outer Space Treaty does not explicitly address the concept of safety zones, leading to uncertainties about their legality. The absence of clear guidelines raises questions about how these zones would be managed and enforced without violating the non-appropriation principles. Hence, establishing safety zones could be seen as a *de facto* claim of territorial sovereignty, as these zones might imply exclusive control over certain areas, thus conflicting with the non-appropriation principle of Article II.

According to Article VI of the Outer Space Treaty, states bear international responsibility for national activities in outer space, including those conducted by non-governmental entities. This means that a state must authorise and continuously supervise the activities of its entities in space, and any non-governmental actor wishing to establish a safety zone must seek authorisation from its respective government. Therefore, safety zones cannot exist independently and require enforcement by a higher authority, such as a state or another recognised entity. This means that safety zones can only be considered to be belonging to private entities if they are protected by the state or other recognised entities. In the context of space law, sovereignty claims are prohibited under Article II of the Outer Space Treaty. Consequently, property rights claims, including establishing safety zones over land in outer space, on the Moon, and other celestial bodies, currently lack legal validity because such claims cannot be recognized or protected by any state or other entity.

Safety zones could exacerbate existing inequalities among space-faring nations. Wealthier and technologically advanced countries might establish extensive safety zones, effectively monopolising certain areas and resources. This could limit opportunities for emerging space nations and create a *de facto* hierarchy in space exploration. Such actions potentially violate the principle of freedom of exploration and use outlined in Article I of the Outer Space Treaty, leading to disputes and tensions among nations. This would undermine the spirit of collaboration and mutual benefit and contradict the principle of peaceful use of outer space. The establishment of safety zones in outer space presents a complex challenge and it must be implemented in a manner that respects the non-appropriation principle while promoting international cooperation. Through transparent, equitable, and inclusive measures, the international community can navigate the controversies surrounding safety zones and ensure the responsible and peaceful use of outer space for all.

To facilitate resource extraction activities in outer space, a legal framework is needed to regulate mining operations for commercial purposes. Without effective and appropriate governance, commercial resource exploitation in space could lead to legal issues such as economic disparities among nations, restricted access to outer space, the Moon, and other celestial bodies, excessive resource exploitation, and environmental damage. This is particularly concerning because private companies or operators prioritise profit and shareholder interests, often neglecting the long-term consequences of their activities. Furthermore, when discussing the extent of liability for non-state actors within the international legal system, particularly within the framework of international space law, it is crucial to understand their liability and responsibility under international law. The focus of non-state actors, including private companies, is often economic interests, which can conflict with broader legal and ethical considerations. Therefore, it is essential to implement robust legal and regulatory frameworks to mitigate risks and ensure that commercial space activities are conducted responsibly and sustainably, as promoted in the Artemis program regarding sustainable space activities.

Private companies must implement measures to prevent environmental damage and preserve the integrity of space environments. Regulatory frameworks and best practices should be established to address the potential risks associated with space mining activities, ensuring that private operations adhere to the standards set forth in the Outer Space Treaty. As the private sector plays an increasingly significant role in space exploration and resource extraction, it is crucial to ensure that these activities align with the principles outlined in Article IX. By promoting international cooperation, preventing harmful contamination, and avoiding interference, the integration of private sector activities in space mining can contribute to the sustainable and equitable use of outer space for the benefit of all humankind.

8.6 Conclusion

It is important to consider the potential issues surrounding deconfliction in outer space, as it could lead to uncertainty about leadership and the risk of pseudo-sovereignty, which would violate Article I of the 1967 OST. Additionally, the lack of legal clarity for space exploration and activities could create significant problems in the future. While initiatives like Artemis could advance space technologies, it's evident that restricting the progress of space technologies is not feasible at this stage. Nevertheless, it's crucial to establish internationally accepted regulations and bodies before undertaking such activities. Bodies such as the International Seabed Authority could serve as a model for preventing excessive claims on the lunar surface by implementing a 'safety zone' through deconfliction measures, as proposed in the Artemis Accords.

References

Aust, A. (2013). *Modern treaty law and practice* (p. 218). Cambridge University Press.

Baca, K. A. (1993). Property rights in outer space. *Journal of Air Law and Commerce, 58*(4), 1049.

Barnes, R. (2009). *Property rights and natural resources* (Studies in International Law, Vol. 22). Oxford: Hart Publishing.

Beaulac, S. (2003). The social power of Bodin's 'sovereignty and international law.' *Melbourne Journal of International Law,* 4.

Bianchi, A., et al. (2015). *Interpretation in international law* (p. 3). Oxford University Press.

Bodin, J. (1992). *On sovereignty: Four chapters from the six books of the republic* (J. H. Franklin, Ed. & Trans.). Cambridge University Press.

Byers, M., & Boley, A. (2023). Who owns outer space? International law, astrophysics, and the sustainable development of space. In *Cambridge Studies in International and Comparative Law*. Cambridge University Press.

Cheng, B. (1965). United Nations resolution on outer space: "Instant" international customary law? *Indian Journal of International Law, 5*(1), 23–48.

Crawford, J. (1994). In J. H. Franklin (Ed.), *The creation of states in international law*. Cambridge University Press.

Deplano, R. (2021). The Artemis Accords: Evolution or revolution in international space law? *International and Comparative Law Quarterly, 70*(3), 799–819. https://doi.org/10.1017/S0020589321000142

Dobos, B. (2022). Tortoise the Titan: Private entities as geoeconomic tools in outer space. *Space Policy, 60*, 1.

Epstein, R. A. (1990). Property and necessity. *Harvard Journal of Law and Public Policy, 13*(2), 3–4.

Fenn, P. T. (1925). Justinian and the freedom of the sea. *American Journal of International Law, 19*(4), 716–727.

Gorove, S. (1968). Interpreting Article II of the Outer Space Treaty. *Proceedings of the Eleventh Colloquium on the Law of Outer Space, 11*, 349–354.

Guilfoyle, D. (2017). Article 92 of LOSC: Status of ships. In A. Prölß (Ed.), *United Nations Convention on the Law of the Sea: A commentary* (p. 700). C.H. Nomos.

Herdegen, M. (2013). Interpretation in international law. In R. Wolfrum (Ed.), *Max Planck encyclopedia of public international law* (p. 123). Oxford University Press.

International Law Commission (ILC). (2018). *Report of the International Law Commission on the work of its seventieth session* (30 April–1 June and 2 July–10 August 2018) (UN Doc A/73/10, p. 23). United Nations.

Lim,J.(2020).*China'sChang'e-5mission:Flaggingthemoonforconflict?*RetrievedJuly22,2024,from https://www.aspistrategist.org.au/chinas-change-5-mission-flagging-the-moon-for-conflict/

Malanczuk, P. (1997). *Akehurst's modern introduction to international law* (7th ed.). Routledge.

McKeown, B., et al. (2022). Artemis Accords: Are safety zones practical for long-term commercial lunar resource utilisation? *Space Policy, 62*, 101504.

Merriam-Webster. (n.d.). *Pseudo*. In Merriam-Webster.com *dictionary*. Retrieved from https://www.merriam-webster.com/dictionary/pseudo

Mills, A. (2023). The privatisation of private (and) international law. *Current Legal Problems, 76*, 75–128.

NASA. (2020). *The Artemis Accords principles for cooperation in the civil exploration and use of the moon, mars, comets and asteroids for peaceful purposes*. Retrieved from https://www.nasa.gov/specials/artemis-accords/img/Artemis-Accords-signed-13Oct2020.pdf

NASA Office of Inspector General. (2011). *NASA's management of moon rocks and other astromaterials loaned for research, education, and public display* (IG-12-007, p. 8). https://www.oversight.gov/sites/default/files/oig-reports/IG-12-007.pdf

O'Connell, D. P. (1984). *The international law of the sea* (Vol. II, p. 800). Clarendon Press.

Office of Space Commerce. (n.d.). *President signs executive order on space resource utilization.* Retrieved April 29, 2024, from https://www.space.commerce.gov/president-signs-executive-order-on-space-resource-utilization/

Pop, V. (2000). Appropriation in outer space: The relationship between land ownership and sovereignty on celestial bodies. *Space Policy, 16*(4), 275–282.

Shirley, M. M. (1992). The what, why, and how of privatization: A World Bank perspective. *Fordham Law Review, 60*(1), 23–36.

Stirn, A. (2020). Do NASA's lunar exploration rules violate space law? *Scientific American.* Retrieved September 7, 2024, from https://www.scientificamerican.com/article/do-nasas-lunar-exploration-rules-violate-space-law

Tapio, J. (2024). Freedom of exploration and use of outer space. In T. Leclerc (Ed.), *Space law: Legal framework for space activities* (pp. 59–82). Wiley.

Tronchetti, F. (2009). *The exploitation of natural resources of the Moon and other celestial bodies: A proposal for a legal regime* (p. 20). Brill | Nijhoff.

U.S. Embassy & Consulates in Italy. (n.d.). *Biden is committed to NASA's Artemis program for the moon and beyond.* Retrieved April 29, 2024, from https://it.usembassy.gov/biden-is-committed-to-nasas-artemis-program-for-the-moon-and-beyond-2/

United Nations. (1967). *Treaty on principles governing the activities of states in the exploration and use of outer space, including the moon and other celestial bodies* (Outer Space Treaty), 610 U.N.T.S. 205. Retrieved from https://www.unoosa.org/oosa/en/ourwork/spacelaw/treaties/outerspacetreaty.html

United Nations. (1969). *Vienna convention on the law of treaties*, 1155 U.N.T.S. 331 (entered into force Jan. 27, 1980). Retrieved from https://treaties.un.org/Pages/showDetails.aspx?objid=080000028007d42f

United Nations. (1979). *Treaty on agreement governing the activities of states on the Moon and other celestial bodies* (G.A. Res. 34/68, U.N. Doc. A/RES/34/68). Reprinted in *18 I.L.M.* 1434. Entered into force on 11 July 1984.

United Nations Office for Outer Space Affairs. (2021). *Guidelines for the long-term sustainability of outer space activities of the Committee on the Peaceful Uses of Outer Space* (ST/SPACE/79). Retrieved from https://www.unoosa.org/oosa/en/oosadoc/data/documents/2021/stspace/stspace79_0.html

Vedda, J. A. (2004). Space commerce. In E. Sadeh (Ed.), *Space politics and policy* (pp. 201–227). Springer.

Chapter 9
Artemis Accords and the Developing Countries: A Catalyst for Degenerative Equality and Mutual Benefit of Space/ Moon Resources

Khafayat Yetunde Olatinwo and Onize Felicity Ifashe

Abstract It would be safe to state that the Outer Space Treaty and the Moon Agreement are filled with relatable and encompassing interest-based provisions. The principles of *res communis* and Common Heritage make sense at the combined reading of Art.4, 6 (2), 7 (b & d), & 8 of the Moon Agreement and Article I of the Outer Space Treaty. This chapter delves into the intention of the world community for international space laws, to which equality and mutual benefit of resources exploited in the pristine environment are paramount irrespective of the level of economic, scientific and technological. The chapter highlights the fact that the interest of mankind taken into consideration in the exploration and exploitation of both space and moon resources accounts for the membership of both Laws. Hence, parties embark on exploitation with the mindset that such activities will and must benefit not only it but also developing countries since space exploration involves huge financial capability. The chapter also discusses international efforts at achieving equality and mutual benefit of space/moon resources, while juxtaposing it with the act of the United States, spare heading another set of Principles for Cooperation in the Civil Exploration and Use of the Moon, Mars, Comets and Asteroid for Peaceful Purposes known as the Artemis Accords. The chapter will address the implications of the provision of the Artemis Accords on the existing international provision of equal benefits of resources; in other words, are developing countries or mankind considered in the making of the Artemis Accords?

Keywords Common heritage · Space exploration · Space/Moon Resources · Developing countries · Technology transfer · Beneficial sharing

---

K. Y. Olatinwo (✉)
Kwara State University, Malete, Nigeria
e-mail: khafayat.olatinwo@kwasu.edu.ng

O. F. Ifashe
Nigerian Law School, Abuja, Nigeria

S. Bhat B., A. Variath (eds.), *Artemis Accords and Resource Mining in Outer Space*, Issues in Space, https://doi.org/10.1007/978-981-95-0152-6_9

9.1 Introduction

In the early years of space exploration, equal benefit of both exploration and most certainly, exploitation and its peaceful use were paramount in the mind of states which was eventually expressed and birthed, firstly, into the principles on the use of space (United Nations General Assembly, 1963a, 1963b), then Outer space Treaty (United Nations, 1967a, b, c) and other space regime (United Nations, 1967a, b, c, 1968, 1972, 1975) including the Moon Agreement (United Nations, 1967a, b, c). Accordingly, the Outer space Treaty emphasis the importance of equal benefits to the extent that "[t]he exploration and use of outer space, including the Moon and other celestial bodies, shall be carried out for the benefit and in the interests of all countries, irrespective of their degree of economic or scientific development, and shall be the province of all mankind. Outer space, including the Moon and other celestial bodies, shall be free for exploration and use by all States without discrimination of any kind, on a basis of equality and in accordance with international law, and there shall be free access to all areas of celestial bodies. There shall be freedom of scientific investigation in outer space, including the Moon and other celestial bodies, and States shall facilitate and encourage international cooperation in such investigation (Art. I)."

To be able to reap the benefit of space resources by individual states requires huge financial investment and other resources like personnel, which may be quite and almost impossible for some developing and underdeveloped countries. Leaving this provision without more is tantamount to a dead law. This, the international community put in place several mechanisms for the peaceful use and equal benefit of space resources such as United Nations Committee on the Peaceful se of Outer Space (United Nations, 1988; European Political Cooperation Documentation Bulletin, 1989; United Nations General Assembly, 1996a, b), the International Telecommunication Union which manages the allotment of slots in space, the encouragement of bilateral relationship to facilitate the sharing of space benefit harnessed by the other State (National Academy of Sciences, n.d.; Diederiks-Verschoor & Kopal, 2008) and at global level, there are worldwide space organisation whose objectives are to advance and foster international cooperation as it relates to the benefit of space resources (International Telecommunication Union [ITU], International Telecommunication Satellite Organisation [ITSO], International Maritime Satellite Organisation [IMSO], n.d.; Olatinwo, 2021).

With all these efforts, it would be strange to see states whose space activities are tantamount to self-interest-based space exploration, one without the intention to benefit humankind. The activity of the US with its Apollo program ought to have been an eye opener and an indication to which the international community should take a drastic measure to safeguard the result of lunar mining. It is worth mentioning here that the United States Apollo program has been used to bring back lunar materials to Earth (NASA, n.d.-a, n.d.-b, n.d.-c, n.d.-d) without sharing with any enterprise-like international body as envisaged by the Moon Agreement, and without any objection from any nation *could be said* to have created an implied approval and precedent to other developed nations that may want to do that. The United States, in asserting ownership of these lunar samples, has been prosecuting cases against people who have stolen from the samples (United States v. One Lucite Ball Containing Lunar Material, 2003; Parvin, 2003). However, the argument that may be put forward by the United States is that the exploitation of space resources for scientific research/purpose is permitted (Moon Agreement, 1979, Art. 6(2)). Another pointer was the attitude of the US in not signing (Editorial, 2020a, b, c) the Moon Agreement, which is deemed as furtherance to the Outer Space Treaty, particularly on exploration and use for humankind. It is therefore obvious, and by championing the Artemis Accords, that the intention for the lunar program may not necessarily be for the benefit of present and future generations. This can be seen in the Accords themselves as there is not one single reference that the exploitation will benefit mankind as preached in the early space regime.

9.2 The Early Regimes and Equal Benefit of Space/ Moon Resources

The exploration of space and the Moon has always been closely linked to the ideals of global collaboration and benefit distribution. This article will explore the early space and lunar resource regimes, emphasizing the underlying treaties and agreements that guarantee the equitable distribution of space advantages to all nations, irrespective of their economic or technological prowess (Outer Space Treaty, Art. I). The Treaty on Principles Governing the Activities of States in the Exploration and Use of Outer Space, including the Moon and Other Celestial Bodies, also referred to as the Outer Space Treaty (Outer Space Treaty), is the cornerstone of international space law. The Outer Space Treaty, enacted in 1967, established the fundamental ideas for space exploration. Hereunder, nations are allowed to explore and use space, including the Moon and other celestial bodies, and any actions conducted there must benefit all of humanity (Outer Space Treaty, Art. II). Its first provision emphasis the intention of the international community to the extent that Outer space and moon resources must utilised for the benefit of all, therefore, "The exploration and use of outer space, including the Moon and other celestial bodies, shall be carried out for the benefit and in the interests of all countries, irrespective of their degree of economic or scientific development, and shall be the province of all mankind' not for one country, not for few countries but for mankind (Outer Space Treaty, Art. I)."

The Outer Space Treaty buttresses those national claims to sovereignty, use, or appropriation of space are not permitted in outer space (Article II). This provision seeks to maintain space as a common property by preventing colonialism based on the use of space. In addition, the treaty stipulates that space exploration and use must be carried out in a way that prevents hazardous pollution and unfavorable environmental impacts. In furtherance to the principles of the Outer Space Treaty, the Agreement Governing the Activities of States on the Moon and Other Celestial Bodies, known as the Moon Agreement, was adopted in 1979 and entered into force in 1984. This treaty further elaborates on the use of lunar resources and emphasizes the need for international cooperation and equitable sharing of benefits (Moon Agreement, 1979, Art. 11). Accordingly, Article 11 of the Moon Agreement states that the Moon and its natural resources are part of humanity's shared heritage. By clearly stipulating that no state or entity may usurp lunar resources, this principle goes beyond the Outer Space Treaty (Moon Agreement, 1979, Art. 11). The agreement calls for the establishment of an international regime to govern the exploitation of these resources, ensuring that activities are conducted for the benefit of all countries, particularly developing nations (Article 11, Paragraph 5).

The Moon Agreement emphasises that "the exploration and use of the Moon shall be the province of all mankind and that it shall be carried out for the benefit and in the interests of all countries, irrespective of their degree of economic or scientific development. It enjoined space actors to pay due attention to 'the interests of present and future generations as well as to the need to promote higher standards of living and conditions of economic and social progress and development in accordance with the Charter of the United Nations (Moon Agreement, 1979, Art. 11)." The legal framework for space exploration and the fair distribution of its benefits have been significantly shaped by the United Nations. The Moon Agreement and the Outer Space Treaty's tenets are reaffirmed in a number of UN General Assembly resolutions. For example, international cooperation should be carried out in a way that benefits all countries, with special attention to the interests of developing nations, according to the declaration on international cooperation in the exploration and use of outer space for the benefit and in the interest of all states, taking particular account of the needs of developing countries (Moon Agreement, 1979, Art. 11(5)). The importance of international cooperation in space operations is further highlighted by UN resolution 68/74, which was adopted in 2013 (Moon Agreement, 1979, Art. 4). It seeks the sharing of scientific information and the promotion of capacity-building in space research and technology, particularly for developing nations. So, if there were in existence provisions which disallow self-interest based exploitation and use of moon resources, why then Artemis Accords.

9.2.1 Challenges and Criticisms

Even with these early regimes' good intentions, there have been many obstacles to overcome before the concepts of equitable benefit-sharing can be applied in practice. The Moon Agreement and the Outer Space Treaty have drawn criticism for not having explicit procedures in place to ensure that space resources are distributed fairly (United Nations General Assembly, 1996a, b). Furthermore, the current legal framework is facing new issues due to the speed at which space technology is developing and the growing participation of commercial organisations in space research. The commercialisation of space resources poses concerns about how to strike a balance between private interests, fair benefit-sharing, and humanity's shared history (United Nations General Assembly, 2013).

9.2.1.1 International Efforts at Equal Benefit for Developing Countries

The international community has been working hard (Howard, 2017) to guarantee that developing and underdeveloped nations benefit equally from space exploitation, building on the fundamental ideals set by early agreements like the Moon Agreement and the Outer Space Treaty. These activities, which vary from capacity-building programs to regulatory important guidelines for global collaboration and the fair distribution of gains, were set by the early space and lunar resource regimes, especially the Moon Agreement and the Outer Space Treaty. These treaties demonstrate a commitment to ensuring that space exploration and utilisation benefit all of humanity and to preventing the monopolisation of space resources. Nonetheless, the dynamic terrain of space exploration demands continuous endeavors to tackle the obstacles and deficiencies within the existing legal structure (Atkins, 2022). Frameworks are all part of an attempt to promote equitable sharing of space technological developments/advancement and inclusive involvement. The role of the United Nations in terms of encouraging the fair use of space resources has been commendable.

The UN has taken the lead in overseeing space exploration and utilisation for the benefit of all people through the United Nations Committee on the Peaceful Uses of Outer Space (UNCOPUOS) (Nelson, 2020) founded in 1959. The UN has assumed the lead in supervising space research and use for the benefit of all people; the United Nations Committee on the Peaceful Uses of Outer Space (UNCOPUOS) was established in 1959 (United Nations Office for Outer Space Affairs, n.d.; United Nations, n.d.; Paxson, 1993). In addition to supervising the execution of international space agreement, UNCOPUOS offers a platform for communication and collaboration amongst its member nations (Committee on the Peaceful Uses of Outer Space, n.d.). The Space2030 Agenda (Committee on the Peaceful Uses of Outer Space, n.d.), which links space operations to the Sustainable Development Goals (SDGs), is one of the major UN programs. The Space2030 Agenda highlights how important space is to be attaining sustainable development, especially for developing nations. It promotes international collaboration on space operations, knowledge transfer, and capacity building to assist developing nations in utilising space technology. Furthermore, UN-SPIDER, the United Nations Platform for Space-based Information for Disaster Management and Emergency Response, assists developing nations in managing disasters through the use of satellite data and technical support. UN-SPIDER helps developing nations reduce the effects of natural disasters, boosting their resilience and sustainable development (Committee on the Peaceful Uses of Outer Space, n.d.).

The International Telecommunication Union (ITU)

One of the most important organisations in ensuring that developing nations have access to satellite communications is the International Telecommunication Union (ITU), a specialised agency of the United Nations (Committee on the Peaceful Uses of Outer Space, n.d.). The ITU assigns global radio spectrum and satellite orbits, which are vital resources for communication and broadcasting services. The ITU guarantees that these resources are used fairly and efficiently through its World Radiocommunication Conferences (WRC), promoting equitable access for all nations, including those with limited technological capabilities. The ITU also offers capacity-building programs and technical assistance to help developing nations enhance their telecommunications infrastructure (United Nations General Assembly, 2007).

Bilateral and Multilateral Agreements

The equitable use of space resources has been greatly aided by bilateral and multinational agreements. Developed nations with sophisticated space programs have collaborated with developing countries to exchange infrastructure, technology, and knowledge.

To encourage the development of space capabilities throughout the continent, the African Union, for example, launched the African Space Policy and Strategy in 2016. The Asia-Pacific Regional Space Agency Forum (APRSAF) serves a similar purpose by promoting collaboration amongst nations in the Asia-Pacific area. APRSAF was founded in 1993 (United Nations General Assembly, 2007) with the goal of advancing space science and technology through collaborative research, information sharing, and capacity building. By promoting regional collaboration, APRSAF assists developing nations in the region in improving their space capabilities and taking an active role in space exploration.

Capacity-Building and Technology Transfer

Building capacity and transferring technology are essential to allowing developing nations to reap the benefits of space exploration. The goal of numerous multinational projects and programs is to improve these countries' infrastructure and technical capabilities. The European Space Agency (ESA) supports developing nations through a number of initiatives, such as the TIGER (United Nations General Assembly, 2007) initiative, which helps African nations manage their water resources by utilising Earth observation data (United Nations General Assembly, 2007). Through training, technical assistance, and satellite data access, ESA assists these nations in gaining the know-how required to utilise space technology for sustainable development. Furthermore, the Emerging Space Leaders Grant Program of the International Astronautical Federation (IAF) fosters capacity-building. Through participation in international space conferences, this initiative enables young professionals and students from underdeveloped nations to network with specialists worldwide and exchange knowledge (European Space Agency, n.d.). Even with these initiatives, there are still a lot of obstacles to overcome before developing nations can receive fair advantages. A lack of infrastructure, a lack of technical ability, and financial constraints prevent many developing countries from engaging in space activities.

In addition, the fast commercialisation of space presents fresh obstacles to fair resource distribution. The international community has advanced the fair use of space resources for the mutual benefit of all nations with notable progress. The UN, bilateral and multinational agreements, and different capacity-building programs have made it possible for developing nations to take part in and profit from space research (European Space Agency, n.d.)

9.3 The U.S. Lunar Program: An Ignored Forewarning

With an even more expansive ambition, the Artemis program embodies the spirit of the Apollo missions and marks a significant advancement in space exploration. But even as this initiative moves the US closer to its lunar goals, it is important to consider the possible drawbacks and cautions of doing such a large-scale project. Although the U.S. Lunar Program holds great potential for scientific and technological progress, it also presents serious questions regarding international space law, the geopolitical environment, and the fair allocation of lunar resources, especially with regard to developing nations.

The United States has long been a pioneer in space exploration, having landed the first humans on the moon in 1969 and constructing the Space Shuttle and the International Space Station(Kimani, 2018). By the end of the decade, the Artemis Program, the twin sister of Apollo and the Greek goddess of the moon, aims to establish a long-term human presence on the moon. The building of the Lunar Gateway, a space station that will circle the Moon, crewed and uncrewed landings, and the use of lunar resources to assist upcoming Mars missions are all part of this program (Zhao, 2016). In 2020, the Artemis Accords were presented as a framework for global collaboration within the Artemis Program. Principles, including emergency aid, interoperability, transparency, and sustainable use of space resources, are outlined in these bilateral agreements. Currently, a number of nations have signed the Artemis Accords, endorsing the American ambition for lunar exploration (NASA, n.d.-a, n.d.-b, n.d.-c, n.d.-d).

Significant concerns concerning international space law, the Outer Space Treaty, and the Moon Agreement are brought up by the Artemis Program and the Accords. The Outer Space Treaty, which was signed in 1967, states that no national government may use any method to appropriate space, including the Moon and other celestial bodies. It states that space exploration should be for the benefit of all people and forbids any country from asserting its sovereignty over space or celestial bodies.

The Artemis Accords reinforce the Outer Space Treaty principles (Hawkins, 2021). They present novel ideas, like creating 'safety zones' to guard against detrimental interference with lunar operations. Opponents (U.S. Department of State, 2023) contend that these safety zones can be seen as a type of territorial claim, which might be in violation of the Outer Space Treaty ban on national appropriation. In addition, the Accords place a strong emphasis on the use of space resources, a subject that continues to be controversial in the international community.

The Moon Agreement (NASA, 2024a, b), which went into effect in 1984, goes a step further and declares the Moon and its resources to be part of the collective heritage of all people. However, only a small number of nations have ratified the accord, indicating its lack of widespread support (NASA, 2024a, b). Conversely, the Moon Agreements' multilateral strategy may be compromised by the bilateral and less widely embraced nature of the Artemis Accords. The United States Lunar Program has enormous geopolitical consequences, with the potential to reshape the balance of power in space exploration. The Artemis Accords seek to unite like-minded nations and promote cooperation amongst those with comparable ideals and goals for space exploration. However, this strategy can result in the creation of exclusive coalitions, which would marginalise the nations that are not signatories to the Accords (Outer Space Treaty, 1967). Art II). China and Russia, both major spacefaring nations, have not signed the Artemis Accords and have expressed concerns about the U.S.-led initiative (Space Policy Review, 2024). Instead, China has been advancing its own lunar program, including plans for a lunar research station and crewed missions.

The potential rivalry between the U.S. and China in space could lead to a new era of space competition reminiscent of the Cold War with significant geopolitical ramifications (Space Policy Review, 2024Art, II). Developing nations can miss out on the scientific and economic benefits of lunar exploration since they frequently lack the financial means and technical know-how to take part. Furthermore, as developing countries are less able to defend their interests in space, the creation of safety zones surrounding lunar missions may result in conflicts over resource-rich areas. As per the Artemis Accords, the U.S. Lunar Program is a noteworthy achievement in space exploration that has the capacity to propel scientific and technical progress (Space Policy Review, 2024). It also acts as a warning, though, about the dangers and difficulties that could arise from lunar research. The necessity for a thorough and inclusive approach to space governance is highlighted by the consequences of international space law, the dangers for developing nations, and the larger geopolitical environment.

9.4 Artemis Accords and Moon Resources

The Artemis Accords, an international framework for space exploration, seek to create standards for the peaceful and sustainable use of space, including the Moon. The extraction and usage of lunar resources is central to the Artemis program, as they hold enormous promise for scientific discovery and economic prosperity. However, this ambitious attempt creates significant legal, ethical, and geopolitical concerns about the management and distribution of these resources. The Artemis Accords are a set of principles guiding the exploration and use of celestial bodies, including peaceful purposes, transparency, interoperability, emergency assistance, and space object registration. These principles ensure international cooperation, promote safety, and promote transparency, ensuring the safety and accountability of space exploration.

9.4.1 Legal Framework for Lunar Resource Utilisation

The use of lunar resources is an important part of the Artemis Accords, but it is also one of the most contentious issues. The Outer Space Treaty, as the fundamental legal document controlling spates activities, including on the Moon and other celestial bodies, is not subject to national appropriation and must be used to benefit all countries. It provides for the peaceful exploration and use of outer space, but it does not provide specific guidelines for the extraction of resources. The gap in the legal framework has led to differing interpretations and debates about the legality of resource utilisation. The Artemis Accords attempt to address this by promoting the idea that space resources can be extracted and used, provided that such activities comply with international law and are conducted transparently and sustainably (China National Space Administration, 2022; The State Council of the People's Republic of China, 2022). In addition to the Outer Space Treaty, the Moon Agreement specifically addresses the exploitation of lunar resources. The Moon Agreement designates the Moon and its resources as the 'common heritage of mankind' and calls for an international regime to govern its use (NASA, n.d.-a, n.d.-b, n.d.-c, n.d.-d). However, the Moon Agreement has not been widely ratified, with only a few countries being parties to it, limiting its impact on the regulation of lunar resource activities (Agreement Governing the Activities of States on the Moon and Other Celestial Bodies [Moon Agreement], 1979, Art. II).

9.4.2 Implications For International Cooperation

The Artemis Accords have the potential to significantly influence international cooperation in space exploration and exploitation. By establishing a set of shared principles, the Accords aim to create a framework for collaborative efforts among signatory nations, which can lead to the pooling of resources, expertise, and technology, enhancing the capabilities of participating countries and accelerating the pace of lunar exploration (Bini, 2008). However, the bilateral nature of the Artemis Accords has also raised concerns about the inclusiveness and fairness of the framework (Wedenig and Nelson, 2023). While the Accords promote cooperation among like-minded nations, they may exclude countries that are not aligned with the U.S.-led initiative, potentially leading to a fragmented approach to space governance. Developing countries, in particular, may find it challenging to participate in lunar resource activities due to financial and technological constraints.

9.5 Equitable and Beneficial Sharing: The Future of Equal Benefit to Developing Countries in the Artemis Accords

A key concern in the discussion surrounding the use of lunar resources is how to distribute the benefits. Both the Moon Agreement and the Outer Space Treaty stress that space exploration and use should benefit all nations, regardless of their level of development. The Artemis Accords support the use of space resources, but they also face the challenge of making sure that the benefits are distributed equally among participating and non-participating nations (Mallick, 2023). In order to address this, the Accords promote international partnerships and the release of scientific data (Deplano, 2021). By providing this, "the Signatories are committed to transparency in the broad dissemination of information regarding their national space policies and space exploration plans in accordance with their national rules and regulations. The Signatories plan to share scientific information resulting from their activities pursuant to these Accords with the public and the international scientific community on a good-faith basis, and consistent with Article XI of the Outer Space Treaty."

Nevertheless, more tangible mechanisms are needed to guarantee that developing nations can benefit from lunar resource use. This could involve international funding establishments to support participation by nations with limited space capabilities, technology transfer initiatives, and capacity-building programs. In terms of creating a framework for the sustainable and peaceful exploration of the Moon and other celestial bodies, the Artemis Accords mark a major advancement. Transparency, interoperability, and the fair use of space resources are goals of the Accords' guiding principles (Editorial, 2020a, b, c). However, careful consideration of legal, ethical, and geopolitical aspects is necessary for the proper implementation of these concepts.

Historically, a few developed countries, most notably the United States, Russia, and more lately, China, have the financial means and capacity to make significant investments in infrastructure, technical advancement, and space missions. On the other hand, developing nations have frequently found it difficult to keep up because of their inadequate infrastructure, financial resources, and technological know-how. With an emphasis on Moon-related projects and beyond, the Artemis Accords were created to promote international cooperation in space exploration. The Accords have drawn criticism for perhaps marginalising developing countries despite their admirable goals (Artemis Accords, n.d.-b, S. 4 & 8).

The combined reading of sections 9 and 10 reveal the confused and selfish nature of the accords while it noted "that the utilization of space resources can benefit humankind by providing critical support for safe and sustainable operation, it further provides that the state parties intend to "use their experience under the Accords to contribute to multilateral efforts to further develop international practices and rules applicable to preserving outer space heritage." This chapter argues for essential reforms to the Artemis Accords in order to defend developing countries' rights and interests while also critically examining the Accords' inherent shortcomings in providing fair benefits to them.

The Artemis Accords have some key flaws as its top-down approach to decision-making prioritises the interests of big spacefaring states. Formal procedures to guarantee developing nations' active involvement in crucial choices impacting space exploration endeavors are absent from the Accords. Due to their lack of financial and technological prowess, these countries' needs and opinions might not be properly taken into account. As a result, the Accords might maintain the disparities in space exploration that now exist, where developing nations are merely observers rather than active players (Artemis Accords, Preamble & Sections 3–5, n.d.-a). Developing countries' ability to influence decisions that directly affect their future chances in space exploration is stifled by this limited involvement. The Artemis Accords are based on the fundamental idea of the equitable use of space resources, but they don't include any explicit rules or procedures to guarantee that developing nations receive an equal share of the moon and other space resources. There is no mention of the program benefiting developing countries in the Accords. In the absence of explicit prohibitions, there is a chance that resource distribution may continue to reward technologically developed countries, widening already existing gaps. It may be difficult for developing nations to acquire or use space resources efficiently, which would undermine the objective of fair benefit (Editorial, 2020a, b, c). Due to the unequal distribution of resources, industrialised countries may monopolise space resources, denying poorer nations the ability to use these resources for their own growth. Although, the Artemis Accords place a strong emphasis on international collaboration (Gross, 2023), they fall short in addressing the necessity of extensive programs for developing nations to enhance their capacity and transfer technology to them. Many developing countries currently lack the significant technological infrastructure and knowledge needed to participate effectively in space exploration (Beldavs, 2018). These nations' ability to advance their space capabilities is hampered by the lack of strong frameworks for technology transfer and capacity building (Artemis Accords, n.d.-c). This lack of assistance feeds a vicious circle of dependency in which developing countries are unable to move toward space exploration self-sufficiency because they are dependent on the generosity and technological breakthroughs of wealthier nations.

There are significant differences in science and economy between industrialised and developing countries. The lack of clear actions to resolve these discrepancies in the Artemis Accords could cause the gap to increase. If poor nations' scientific and technical progress is not adequately supported by specific programs, the Accords may fall short of guaranteeing that space exploration serves the interests of all countries equitably (SpaceHubs Africa, 2024; Editorial, 2023). These differences may lead to a technology divide that further marginalises the contributions and advantages of space exploration made by poor countries, making it impossible for them to cooperate or compete on an equal basis with their more developed counterparts. The structure of the Artemis Accords runs the potential of encouraging a kind of neocolonialism in space, in which more developed nations control access to and dominate space operations and resources. Due to a lack of infrastructure and technology, developing nations might be forced to take a secondary role and not be able to fully participate in the advantages of space travel. This dynamic is reminiscent of past exploitation patterns in which wealthy and powerful nations take resources and wealth from less developed areas, leaving them dependent and impoverished (NASA, 2024a, b). Such exploitation in space presents moral questions and emphasises the need for more just and inclusive government structures.

Developing countries may also be disproportionately affected by the environmental impacts of space activities. The Artemis Accords do not provide comprehensive guidelines for mitigating the environmental risks associated with space exploration, such as space debris and lunar surface disruption. These issues could have long-term implications for all nations, but developing countries may lack the resources to address or mitigate these risks effectively (NASA, 2024a, b). Furthermore, the legal frameworks governing space activities are still evolving, and without active participation from developing countries, their interests may not be adequately represented in international space law.

9.6 Path Forward

The Artemis Accords have to be revised to take into account the following factors in order to remedy these flaws and guarantee that impoverished nations can profit from space exploration. The Accords ought to create official channels for including poor nations in space exploration decision-making. This can entail establishing committees or forums with a specific focus to guarantee varied participation and opinions from all participating countries. The Accords can more effectively address the concerns of developing nations and guarantee more fair participation by taking their opinions into account (Beldavs, 2018). This can be achieved by redrafting the Accords to reflect the customary intention of the international community as enshrined and reflected in the early space regimes to the effect that resources harnessed under the program would be used, having in mind the developing countries. Clear criteria must be developed and put into effect for the fair allocation of space resources. These recommendations must incorporate particular clauses aimed at guaranteeing developing nations' access to and utilisation of space resources. These recommendations must incorporate particular clauses aimed at guaranteeing developing nations' access to and utilisation of space resources. Enacting procedures to rectify any disparities in the distribution of resources will aid in accomplishing the objective of just compensation (Beldavs, 2018). The Artemis Accords must incorporate comprehensive capacity-building initiatives. The development of infrastructure, technological transfer, and training for developing nations should be the main focuses of these initiatives. Through the advancement of technological capabilities and infrastructure, the Accords have the potential to enable a wider range of people to participate in space exploration.

The Artemis Accords, which presumably seek to promote international cooperation and guarantee that the advantages of moon and space activities are distributed fairly among all nations, represent a turning point in space exploration. But achieving these objectives calls for a thorough and well thought out strategy, as a vital component of the Artemis Accords is the requirement for strong international cooperation. The Artemis Accords need to develop from bilateral agreements into a more comprehensive multilateral framework to address this. Engaging global institutions like the United Nations Office for Outer Space Affairs (UNOOSA), which can offer a forum for broader involvement and discourse, can help achieve this. Making inclusive decisions is another essential element. Developing nations need to be represented in talks and decisions about space governance. Working groups, advisory boards, and worldwide forums that make sure all opinions are heard and taken into account can help promote this inclusivity. Utilising public-private partnerships can also hasten resource mobilisation and technological improvements, making it easier to transfer money, expertise, and technology to emerging nations.

Developing countries must build their infrastructure and capabilities in order to participate actively in space activities. Developed countries and international organisations should be willing to offer financial and technical assistance to poorer countries so they can construct their space infrastructure. Grants, low-interest loans, and capacity-building initiatives aimed at advancing knowledge in space science and technology are among possible forms of this assistance. Moreover, educational programs are essential. Creating training and educational programs in space science and technology can assist developing nations in building a workforce with the necessary skills. Developing nations should ensure that they have the human resources required to participate in space exploration by fostering talent and expertise through international scholarships, exchange programs, and cooperative research efforts.

Furthermore, promoting the creation of national space agencies in underdeveloped nations can offer a methodical approach to overseeing and arranging space operations. By acting as focal points for global cooperation and policy execution, these organisations may make sure that poor nations are well-positioned to gain from international space projects (Gross, 2023). Creating strong legal and regulatory frameworks is essential to guaranteeing that space operations are carried out sustainably and ethically. It is important to assist developing nations in creating and bringing their national space legislation into compliance with international standards. This can address particular national interests and goals while guaranteeing adherence to international treaties and agreements, such as the Moon Agreement and the Outer Space Treaty (Matusiak, 2023).

In consultation with all stakeholders, precise norms and standards for the ethical and sustainable use of lunar and space resources must be devised. These rules ought to guarantee that emerging nations' interests are fairly reflected and safeguarded. Building confidence and guaranteeing adherence to international agreements also depend on encouraging accountability and transparency in space activity.

Mechanisms for monitoring and reporting on space activities should be established to promote accountability and prevent conflicts (Gibbs and Pryke, 1997). Mechanisms for benefit-sharing must be developed in order to guarantee that the advantages of space exploration are dispersed fairly. These can be international funds or revenue-sharing arrangements that designate a percentage of the proceeds from the exploitation of space resources for the benefit of developing nations' sustainable development initiatives. Encouraging the dissemination of scientific data gathered from space missions can be advantageous to all nations. The infrastructure and resources required for developing nations to access and use this data for their purposes in scientific research and development should be made available to them (Chen & Ahmad, 2016).

A further significant approach is to encourage involvement in multinational space missions. It is important to encourage developing nations to take part in cooperative missions, programs, and alliances that offer chances for information sharing and capacity building (United Nations Office for Outer Space Affairs, 2021). By participating, developing nations can make sure they don't fall behind in the quickly rising field of space exploration. Sustainability and ethical considerations are paramount in space exploration (United Nations Office for Outer Space Affairs, 2021). The extraction and utilisation of space resources must be conducted sustainably to avoid (United Nations General Assembly, 2023) depleting or damaging the lunar environment (Tepper, 2022). Ensuring that these operations are carried out properly can be facilitated by the development of international standards for sustainable resource management. Space exploration must also be guided by ethical standards that preserve humanity's shared history and make sure that space activities do not worsen already-existing injustices or give rise to new kinds of exploitation.

9.7 Conclusion

An important step towards a new era of space exploration that 'prioritises international cooperation, transparency, and equitable benefit-sharing' is represented by the Artemis Accords, Even though the Accords lay forth lofty ideals and objectives, careful consideration of the particular difficulties experienced by developing nations is necessary for the successful implementation of these aims and principles (Fleming et al., 2023).

It is vital to augment global cooperation via multilateral involvement and inclusive decision-making procedures (Fleming et al., 2023). Through the provision of a platform for active participation in space governance, the international community can guarantee a more equitable distribution of the advantages associated with space exploration (Fleming et al., 2023). Strong legal and regulatory frameworks are essential for directing the conscientious and long-term use of space resources (Fleming et al., 2023).

Mechanisms for equitable benefit-sharing are also essential. International financing, agreements for revenue sharing, and unrestricted access to scientific data can guarantee that the knowledge and earnings from space exploration support the sustainable development of all countries and encourage foreign missions to advance capacity building and exchange of information.

It is critical to address ethical and sustainable issues. International norms for sustainable resource management must be created by the international community, and space activities must be governed by morality and respect for humanity's shared history. To sum up, the Artemis Accords present a viable structure for the equitable and inclusive use of space resources. We can create a more just and inclusive future for space exploration by strategically working to improve cooperation, increase capacity, create strong legal frameworks, guarantee equitable benefit-sharing, and address environmental and ethical issues. Space exploration has enormous potential, and with these efforts, we can make sure that all countries, especially poor ones, may take part in and profit from this fascinating frontier (International Space Exploration Coordination Group [ISECG], 2013).

References

Agreement Governing the Activities of States on the Moon and Other Celestial Bodies (Moon Agreement), 1979, Art. II. Art. I.

Artemis Accords. (n.d.-a). Preamble & Sections 3–5.

Artemis Accords. (n.d.-b). Sections 4 & 8.

Artemis Accords. (n.d.-c). Sections 2, 3, 7, 8, 12 & 13. NASA.

Atkins, S. (2022, June). *The commercialisation of outer space: How an international securities framework can be the launching pad for a global space economy*. Norton Rose Fulbright. https://www.nortonrosefulbright.com/en/knowledge/publications/1021426e/the-commercialisation-of-outer-space

Beldavs, V. (2018, March 26). *Equitable sharing of benefits of space resources*. The Space Review. https://thespacereview.com/article/4039/1.

Bini, A. (2008). *The Moon Agreement: Its effectiveness in the 21st century* (ESPI Perspectives No. 14). Available at https://www.files.ethz.ch/isn/124689/espi_%20perspectives_14.pdf. Last visited, 2 October 2024.

Chen, K.-W., & Ahmad, T. (2016). Promotion for development of national space legislation in developing states to ensure global space governance. *Presented at the 10th United Nations Workshop on Space Law: Contribution of Space Law and Policy to Space Governance and Space Security in the 21st Century*. United Nations Office for Outer Space Affairs. https://www.unoosa.org/pdf/SLW2016/Panel5/2._Chen_Ahmad_National_space_Legislation_Presentation_Chen.pdf.

China National Space Administration. (2022, January 28). *China's space program: A 2021 perspective*. Retrieved August 14, 2024, from http://www.cnsa.gov.cn/english/n6465652/n6465653/index.html

Committee on the Peaceful Uses of Outer Space. (n.d.). *Committee on the peaceful uses of outer space*. United Nations Office for Outer Space Affairs. https://www.unoosa.org/oosa/en/ourwork/copuos/index.html

Deplano, R. (2021). The Artemis Accords: Evolution or revolution in international space law? *International & Comparative Law Quarterly*, *70*(4), 788–819. Available at https://www.cambridge.org/core/journals/international-and-comparative-law-quarterly/article/artemis-accords-evolution-or-revolution-in-international-space-law/DC08E6D42F7D5A971067E6A1BA442DF1. Last visited, 23 September 2024.

Diederiks-Verschoor, I. H. P., & Kopal, V. (2008). *An introduction to space law* (3rd ed.). Kluwer Law International.

Editorial. (2020a, October 19). Artemis Accords: Why many countries are refusing to sign Moon exploration agreement. *The Conversation*. https://theconversation.com/Artemis-Accords-why-many-countries-are-refusing-to-sign-Moon-exploration-agreement-148134 (Accessed September 18, 2024).

Editorial. (2020b, October 19). *Artemis Accords: Why many countries are refusing to sign Moon exploration agreement.* The Conversation. Available at https://theconversation.com/artemis-accords-why-many-countries-are-refusing-to-sign-moon-exploration-agreement-148134. Last visited 23 September 2024.

Editorial. (2020c, October 19). *Artemis Accords: Why many countries are refusing to sign Moon exploration agreement.* The Conversation. https://theconversation.com/artemis-accords-why-many-countries-are-refusing-to-sign-moon-exploration-agreement-148134#:~:text=Eight%20countries%20have%20signed%20the%20Artemis%20Accords,%20a.

Editorial. (2023, June 20). *Space exploration: How developing countries are now fueling the booming space industry*. Airforce Technology. https://www.airforce-technology.com/sponsored/space-exploration-how-developing-countries-are-now-fuelling-the-booming-space-industry/#:~:text=In%20the%20past%20five%20years%20alone%2C%20Ethiopia%2C%20Mauritius%2C%20have%20all%20put%20their%20first%20satellites%20into%20orbit

European Political Cooperation Documentation Bulletin. (1989). *European political cooperation documentation bulletin* (Vol. 4, No. 2). Archive of European Integration. http://aei.pitt.edu/37316

European Space Agency. (n.d.). *ESA facts*. European Space Agency. https://www.esa.int/About_Us/Corporate_news/ESA_facts

Fleming, M., Lange, I., Shojaeinia, S., & Stuermer, M. (2023). Mining in space could spur sustainable growth. *Proceedings of the National Academy of Sciences, 120*(43), e2221345120. https://doi.org/10.1073/pnas.2221345120

Gibbs, G., & Pryke, I. (1997). International cooperation in space – Developing new approaches. *ESA Bulletin, 89*. European Space Agency. http://www.esa.int/esapub/bulletin/bullet89/gibbs89.htm

Gross, M. (2023, January 27). *The Artemis Accords: International cooperation in the era of space exploration*. Harvard International Review. https://hir.harvard.edu/the-artemis-accords/

Hawkins, L. (2021). *Artemis and Human Landing System program overview*. NASA. Retrieved September 23, 2024, from https://ntrs.nasa.gov/api/citations/20210016717/downloads/21%206%208%20Hawkins%20JANNAF_Final.pdf

Howard, J. (2017, August 28). *Notes on the Moon Agreement*. James Howard. https://jameshoward.us/2017/08.28/comments-moon-treaty/#~:text=The%20moon%20treaty%20may%20be%20a.

International Space Exploration Coordination Group (ISECG). (2013, September). *Benefits stemming from space exploration*. NASA. https://www.nasa.gov/wp-content/uploads/2015/01/Benefits-Stemming-from-Space-Exploration-2013-TAGGED.pdf

International Telecommunication Union (ITU), International Telecommunication Satellite Organisation (ITSO), International Maritime Satellite Organisation (IMSO). (n.d.).

Kimani, J. (2018). *Capacity building in space technology for developing countries: Kenya's perspective*. United Nations Office for Outer Space Affairs. https://www.unoosa.org/documents/pdf/hlf/HLF2018/Pres/6_Kimani_Bonn_Capacity_Building_for_Developing_Countries.pdf

Mallick, S. (2023, September 21). *The Artemis Accords: Changing the narrative from space race to space cooperation.* Space News. Available at https://spacenews.com/the-artemis-accords-changing-the-narrative-from-space-race-to-space-cooperation/#:~:text=The%20Accords,%20announced%20in%202020,%20are%20an%20international. Last visited, 2 October 2024

Matusiak, M. (2023). Artemis Accords – A new era of space law or an initiative of the world's superpower? *Studia Iuridica, 97*, 168–184. https://studiaiuridica.pl/article/538955/en

Moon Agreement, Art. 6(2), Dec. 5, 1979, 18 ILM 1434.

NASA. (2024a). *The Artemis Accords: Principles for cooperation in the civil exploration and use of the Moon, Mars, comets, and asteroids for peaceful purposes*. Retrieved August 2, 2024, from https://www.nasa.gov/specials/artemis-accords/index.html

NASA. (2024b, August 14). *Artemis Accords: Principles for cooperation in the civil exploration and use of the Moon, Mars, comets, and asteroids for peaceful purposes*. https://www.nasa.gov/specials/artemis-accords/index.html

NASA. (n.d.-a). *Apollo 11 mission overview*. https://www.nasa.gov/history/apollo-11-mission-overview (Accessed September 17, 2024).

NASA. (n.d.-b). *Apollo program information*. National Space Science Data Center. https://nssdc.gsfc.nasa.gov/planetary/lunar/Apollo.html (Accessed September 17, 2024).

NASA. (n.d.-c). *Artemis Accords: Principles for cooperation in the civil exploration and use of the Moon, Mars, comets, and asteroids for peaceful purposes*. Retrieved August 14, 2024, from https://www.nasa.gov/specials/artemis-accords/index.html

NASA. (n.d.-d). *Human spaceflight*. NASA. Retrieved September 23, 2024, from https://www.nasa.gov/specials/60counting/spaceflight.html

National Academy of Sciences. (n.d.). *The International Geophysical Year*. http://www.nas.edu/history/igy/ (Accessed August 13, 2024).

Nelson, J. W. (2020). *The Artemis Accords and the future of international space law. Insights, 24*(31). American Society of International Law. https://www.asil.org/sites/default/files/ASIL_Insights_2020_V24_I31.pdf

Olatinwo, K. (2021). Space resources and developing nations: An innuendo for parallel benefit. *Kwara State University Law Journal, 2*(1) http://kwasustaff.kwasu.edu.ng/public/viewpub/260?page=13 (Accessed June 13, 2024)

Outer Space Treaty, (1967). Art II.

Outer Space Treaty, Art. I.

Parvin, C. (2003). *United States v. One Lucite Ball Containing Lunar Material: United States District Court for the Southern District of Florida, 252 F. Supp. 2d 1367*. Quimbee. https://quimbee.com/cases/united-states-v-one-lucite-ball-containing-lunar-material

Paxson, E. W., III. (1993). Sharing the benefits of outer space exploration: Space law and economic development. *Michigan Journal of International Law, 14*(3), 508–513. https://repository.law.umich.edu/mjil/vol14/iss3/

Space Policy Review. (2024). *The Outer Space Treaty and the Moon Agreement: Complementary or contradictory?* Retrieved August 2, 2024, from https://www.spacepolicyreview.com/articles/ost-moon-agreement-complementary-contradictory

SpaceHubs Africa. (2024, August 19). *Regions*. https://spacehubs.africa/regions.

Tepper, E. (2022). The big bang of space governance: Towards polycentric governance of space activities. *International Law and Politics, 54*(485).

The State Council of the People's Republic of China. (2022, January 28). *China's space program: A 2021 perspective*. Retrieved September 30, 2024, from https://english.www.gov.cn/archive/whitepaper/202201/28/content_WS61f35b3dc6d09c94e48a467a.html

U.S. Department of State. (2023). *Space unites us: Artemis Accords—Principles for cooperation in the civil exploration and use of the Moon, Mars, comets, and asteroids for peaceful purposes*. Retrieved September 23, 2024, from https://www.state.gov/artemis-accords/

United Nations. (1967a, January 27). *Treaty on principles governing the activities of states in the exploration and use of outer space, including the moon and other celestial bodies* (610 U.N.T.S. 205, 18 U.S.T. 2410, T.I.A.S. No. 6347, 6 I.L.M. 386). https://treaties.un.org/doc/Publication/UNTS/Volume%20610/v610.pdf (Entered into force on October 10, 1967). [Outer Space Treaty].

United Nations. (1967b, January 27). *Treaty on principles governing the activities of states in the exploration and use of outer space, including the moon and other celestial bodies* (Outer Space Treaty, General Assembly Resolution 2222 [XXI]). https://www.unoosa.org/oosa/en/ourwork/spacelaw/treaties/outer-space-treaty.html (Entered into force on October 10, 1967).

United Nations. (1967c, January 27). *Treaty on principles governing the activities of states in the exploration and use of outer space, including the Moon and other celestial bodies* (the "Outer Space Treaty"). Resolution 2222 (XXI), General Assembly. https://www.unoosa.org/oosa/en/ourwork/spacelaw/treaties/outerspacetreaty.html

United Nations. (1968, April 22). *Agreement on the rescue of astronauts, the return of astronauts and the return of objects launched into outer space* (Rescue Agreement, General Assembly Resolution 2345 [XXII]). https://www.unoosa.org/oosa/en/ourwork/spacelaw/treaties/rescue-agreement.html (Entered into force on December 3, 1968).

United Nations. (1972, March 29). *Convention on international liability for damage caused by space objects* (Liability Convention, General Assembly Resolution 2777 [XXVI]). https://www.unoosa.org/oosa/en/ourwork/spacelaw/treaties/liability-convention.html (Entered into force on September 1, 1972).

United Nations. (1975, January 14). *Convention on registration of objects launched into outer space* (Registration Convention, General Assembly Resolution 3235 [XXIX]). https://www.unoosa.org/oosa/en/ourwork/spacelaw/treaties/registration-convention.html (Entered into force on September 15, 1976).

United Nations. (1988, April 8). *Report of the Committee on the Peaceful Uses of Outer Space* (UN Doc. A/AC.105/411). United Nations.

United Nations. (n.d.). *Application of space to the sustainable development goals*. https://www.un.org/sites/un2.un.org/files/outer-space-and-the-sdgs-en.pdf

United Nations General Assembly. (1963a, December 13). *Declaration of legal principles governing the activities of states in the exploration and use of outer space* (Resolution 1962 [XVIII]). United Nations. https://www.un.org/en/ga/search/view_doc.asp?symbol=A/RES/1962(XVIII)

United Nations General Assembly. (1963b, December 13). *Resolution 1884 (XVIII)*. United Nations. https://www.un.org/en/ga/search/view_doc.asp?symbol=A/RES/1884(XVIII)

United Nations General Assembly. (1996a). *Declaration on international cooperation in the exploration and use of outer space for the benefit and in the interest of all states, taking into particular account the needs of developing countries*, UN Resolution 51/122. https://www.unoosa.org/oosa/oosadoc/data/resolutions/1996/general_assembly_51st_session/ares51122.html#:~:text=UN%20Conference%20on%20Space%20Law%20and%20Policy.%20First.

United Nations General Assembly. (1996b, December 13). *Resolution adopted by the General Assembly on 51/122: Declaration on international cooperation in the exploration and use of outer space for the benefit and in the interest of all States* (UN Doc. A/RES/51/122). http://www.unoosa.org

United Nations General Assembly. (2007, January 15). *Disaster management platform for space-based information for disaster management and emergency response 2006* (U.N. Doc. A/RES/61/110). United Nations Office for Outer Space Affairs. https://www.unoosa.org/oosa/en/ourwork/topics/disaster-management.html

United Nations General Assembly. (2013). *International cooperation in the peaceful uses of outer space*, UN Doc. A/RES/68/74 (GA/11469). https://www.un.org/en/ga/68/resolutions.shtml

United Nations General Assembly. (2023, October 24). *Outer space must be a place for peace and cooperation, not an arms race, speakers affirm, as Fourth Committee takes up space matters* (GA/SPD/788). https://press.un.org/en/2023/gaspd788.doc.htm

United Nations Office for Outer Space Affairs. (2021). *UN guidelines for the long-term sustainability of outer space activities of the Committee on the Peaceful Uses of Outer Space*. United Nations. https://www.unoosa.org/documents/pdf/PromotingSpaceSustainability/Publication_Final_English_June2021.pdf.

United Nations Office for Outer Space Affairs. (n.d.). *Benefits of space for humankind*. United Nations Office for Outer Space Affairs. https://www.unoosa.org/oosa/en/benefits-of-space/benefits.html.

United States of America v. One Lucite Ball Containing Lunar Material, 252 F. Supp. 2d 1367, 16 Fla. L. Weekly Fed. D342 (S.D. Fla. 2003).

Wedenig, S.-M., & Nelson, J. W. (2023, January 26). *The Moon Agreement: Hanging by a thread?* Institute of Air & Space Law. Available at www.mcgill.ca/iasl/article/moon-agreement-hanging-thread#:~:text=As%20such,%20and%20similar%20to%20the%20other%20space. Last visited, 2 October 2024.

Zhao, Y. (2016). The role of bilateral and multilateral agreements in international space cooperation. *Space Policy, 36*, 12–18. https://doi.org/10.1016/j.spacepol.2016.03.002

Chapter 10
India's Tryst with the Goddess of the Moon: The Artemis Accord and Beyond

Mohd Sikandar

Abstract The Artemis Accords serve as a watershed moment in space exploration and cooperation. Coming at a time when more nations than ever are gearing up for space adventure, the Accords attempt to plug the gap left open by the Moon Agreement. It sort of creates a global regime of outer space resource exploration and extraction. However, hidden in the plain language is the ultimate goal of commercial resource exploitation and territorial appropriation leading to what many scholars have called as 'Space Colonisation'. Like its predecessors, the US is attempting to take the benefit of first-mover advantage. However, it cannot do so until it has secured some amount of global acceptance. The Accords are a way of doing that. In this regard, India's signing of the Accords raises some crucial questions. Drafted by the US, the Accords were never negotiated by India. Therefore, to what extent they cater to the needs of the Indian space industry remains to be seen. Secondly, India does not have a developed private space industry like its US counterpart. It raises doubts about India's ability to avail private resource exploitation benefits under the Artemis Accords. Thirdly, the provision of 'safety zones' in the agreement runs counter to India's long-standing position of non-appropriation of celestial bodies. These issues, amongst others, question India's decision to sign the Artemis Accords. The Chapter shall highlight the inequitable relationship that the Accords create that disproportionately benefits US interests with little benefits for India.

Keywords Artemis Accords · Moon Agreement · US Space Programme · India · Privatisation · ILRS

M. Sikandar (✉)
National Law University Odisha, Cuttack, Odisha, India
e-mail: sikandar@nluo.ac.in

S. Bhat B., A. Variath (eds.), *Artemis Accords and Resource Mining in Outer Space*, Issues in Space, https://doi.org/10.1007/978-981-95-0152-6_10

10.1 Introduction

Since time immemorial, humans have been mesmerised by nature's beauty and its resources and in the quest to own nature's gifts, humans have ventured into new worlds, some known and many unknown. Whether it was the discovery of America by Christopher Columbus or the discovery of India by Vasco De Gama, the inherent exploratory nature of humankind has encouraged people to discover new lands and build new colonies. Associated with that nature is the inalienable urge to own and control land and other resources (Barnes, 2009). This cannot be truer for the ever-developing Western civilisation. Simon Winchester (2021) gives a vivid description of the evolving nature of the human-property relationship and man's quest to amass land and its resources. As human wants are insatiable, the earth with its limited resources is now proving to be not enough. As human pressure increases on the planet, accompanied by a consumeristic lifestyle, the earth's resources are bound to be exhausted (Meadows & Randers, 2012). Already, there is a nasty global race ongoing to appropriate what is left on the planet (Klare, 2012). A natural step would obviously be then to look for worlds beyond the world.

Like continental discoveries, the idea of space exploration is not a new one. Humans have always wondered what lies in the space above. The infinite outer space and all its celestial bodies have intrigued humans since the origin of human life itself (Holton & Brush, 2001). However, unlike the dream of space exploration, the idea of outer space resource exploration and extraction is a new one and has gained traction only recently (Pascale, 2023). An interesting development in the arena of the space economy has been the commercialisation and privatisation of the space industry. The advancement and proliferation of space technology, coupled with the development of efficient space transportation systems, has enabled many private players to experiment with the idea of space mining and colonisation (Akpambang & Akanle, 2023). In this respect, Asteroids and the Moon stand as the most potential sites for outer space resource mining. They have been found to be extremely rich in precious minerals like platinum, Helium-3 and water (Feichtner, 2019).

But, in the absence of a universal legal framework governing outer space resource mining, associated property rights on celestial bodies become a contentious issue (Akpambang & Akanle, 2023). The present international framework in the form of the Outer Space Treaty and the Moon Agreement provides us with little guidance with respect to mining in outer space (Zell, 2006). On the contrary, these treaties are built on the idea of outer space being the "common heritage of mankind," thereby denying any property rights in outer space. In short, the absence of an international framework governing outer space resource mining acts as an obstacle to outer space resource extraction and the commercialisation of the space economy.

To do away with this obstacle, countries like the United States have come up with unilateral yet universal legal frameworks in the form of Artemis Accords (Deplano, 2021). These Accords give practical shape to the idea of outer space resource mining and extraction (Kostenko, 2021). By inviting countries to join hands in commercially exploiting celestial bodies, the Accords attempt at rewriting legal principles governing outer space resource appropriation and property rights (Prema et al., 2024). The contention of the drafters that the Accords are in consonance with the established legal principles does not appear to be convincing, especially when read with provisions providing for 'Safety Zones' and extraction and utilization of space resources. In the backdrop of the Moon Agreement, which categorically bars any surface or subsurface resource extraction on the moon (von der Dunk, 2023) and the 'non-appropriation' principle set out in the Outer Space Treaty, the Accords create a parallel and opposite regime for outer space exploration and resource mining.

Amidst such glaring contradictions, the signing of Accords by other countries and by India in particular raises some crucial questions. These questions pertain not only to India's international obligations but also to its ability to utilise the agreement to its benefit. The present chapter, as such, attempts to analyse these questions in a detailed and nuanced manner. In this respect, the chapter has been divided into the following parts. Part II looks into the present international space regime, particularly with respect to outer space resource extraction and the appropriation of celestial bodies. Part III analyses the provisions of the Artemis Accords against the backdrop of the established international legal principles governing outer space resource extraction and exploitation. Part IV discusses the nature and evolution of India's space programme. Part V appraises the signing of Accords by India against its international commitments and technological competency. Part VI gives the conclusion as to the chapter.

10.2 International Space Regime: No Commercial Licenses, Please

The successful launch of Sputnik-I on October fourth 1957, was a turning point in human history. Not only have humans for the first time crossed the earth-space frontier, but it also opened doors for possible problems like the weaponization of outer space, appropriation of celestial bodies and conflict over property rights (Connolly & Bennett, 2022). Thus, the crossing of human activity into a new dimension called for setting rules for beyond the Earth. It was not long before the first international framework governing human activities in outer space were drafted and adopted. With remarkable speed, the international community came up with the first treaty on space exploration and expeditions – the Treaty on Principles Governing the Activities of States in the Exploration and Use of Outer Space, including the Moon and other Celestial Bodies, 1967 or the Outer Space Treaty.

Often termed as the Magna Carta of international space law (Qizhi, 1997), it lays down customary principles governing space exploration and utilisation. One of the cardinal principles enshrined under the treaty states that outer space shall not be subjected to appropriation by claim of sovereignty, by means of use or occupation or by any other means. In other words, the treaty lays down a global norm of 'non-appropriation' of outer space, including the moon and other celestial bodies. In this respect, the bar to national appropriation has to be interpreted in a wider sense, including bar to both sovereign and property rights. As such, states are barred from establishing proprietary links with regard to outer space bodies and objects (Masson-Zwaan & Hobe, 2010; Hasin, 2022). In other words, the non-appropriation principle extends to space resources, including surface and subsurface resources. Further, while the treaty provides for the 'use' of outer space by all nations under Article 1, implied in the language is the in-situ resource utilisation. This is discernible from the language of the provision itself, which declares space and space resources a province of all mankind. The notion that it includes the utilisation of space resources on earth is difficult to accept given the fact that the drafters of the treaty drafted it in terms of the then technological developments (Osada, 2022). In any way, equating 'use' with commercial use would be a far-fetched interpretation of the treaty, an interpretation not in line with the Vienna Convention on the Law of Treaties, 1969 (VCLT) (De Man, 2017).

Besides the Outer Space Treaty, 1967, four other international agreements govern outer space activities: the Rescue Agreement, 1968; the Liability Convention, 1972; the Registration Convention, 1975; and the Moon Agreement, 1979. Among the four international agreements, the Moon Agreement has the most salient role in space governance. At the outset, the Moon Agreement attempted to fill the gaps left open by the Outer Space Treaty (de Zwart, 2021). The incessant lunar exploration missions involving the carrying of lunar resources back home formed the background of the need to regulate lunar explorations and resource exploitation. Although the commercialise lunar explorations, such a possibility was always on the horizon (Hobe & De Man, 2017). As such, the Moon Agreement was drafted in an attempt to prohibit property rights on the moon (Nelson, 2010).

Serving as the complementary document, the agreement reiterates the principle of non-appropriation enshrined in the Outer Space Treaty. In particular, Article 11 of the Agreement removes any ambiguities left by the OST by declaring the Moon, including all celestial bodies and their resources, the 'common heritage of mankind' (Coffeyt, 2009). In effect, the agreement lays to rest any ambiguity relating to the subsurface exploitation of lunar resources. In no uncertain terms, it lays down that no part of the moon, including surface or subsurface resources, shall become the property of any governmental or non-governmental entity, including natural persons. However, the agreement in no way completely bars the exploitation of lunar resources. Evidently, the agreement lays down twin conditions for resource exploitation: the creation of an international regulatory regime and the orderly development and equitable sharing of resources. Read together, the conditions imply coordinated resource exploration and exploitation for scientific investigation and knowledge sharing (Raclin, 1986; Osada, 2022). The international space regime, so to say, is not in favour of granting commercial space licenses.

Notably, the explicit prohibition on property rights and the declaration of celestial bodies as the 'common heritage of mankind' have been the major reasons for the rejection of the Moon Agreement by states. Unlike its predecessor (OST), which has 112 ratifications, the Moon Agreement has been signed only by a handful of 18 countries (*United Nations Treaty Collection*). It is noteworthy that while space-faring nations like the United States and India have signed the OST, they have deliberately ignored the Moon Agreement. In the face of little interest by the international community, no new space treaties have been adopted by countries since the Moon Agreement. Interestingly, the international space law regime has witnessed a shift towards 'soft law' for space governance and exploration. The rigid and binding nature of international treaties has discouraged many space-faring nations from committing to them and thereby getting bound by global norms. On the other hand, soft law instruments have been adopted by nations to operate in space with flexibility and manage their space endeavours, and also contribute to the progressive development of space law (Brunner & Soucek, 2012). As such, the traditional treaty method of space governance has been replaced, to a large extent, by soft law instruments like the United Nations General Assembly Resolutions and country agreements. The United States' Artemis Accords is one such country's soft law instrument for space governance (Neef, 2021). Additionally, in order to fill the gaps in international space law, nations have come up with domestic legislation to cater to their domestic interest. Thus, countries like the United States, Luxembourg, the United Arab Emirates and Japan have passed domestic legislation allowing space mining and commercialisation of the space economy (de Zwart et al., 2023). These agreements and legislations have time and again raised the issue of alleged violation of international law. The following section shall analyse this issue from the perspective of the United States Space Programme.

10.3 In Unilateral Interpretation, We Believe: The United States Space Programme

The launch of Sputnik-I by the former USSR, although shocking, came as a blessing in disguise for the United States. It laid the foundation for a robust space law in the country – the National Aeronautics and Space (NAS) Act of 1958. The answer to the Soviet Union's threatening posture was given in the form of the establishment of a civilian space programme (Gabrynowicz, 2010). In this respect, the NAS Act was instrumental in promoting the idea of "peaceful uses of outer space" in international space law. Since then, the United States space programme has added a new dimension to its civil and military components – the commercialisation and privatisation of space missions. Beginning with the 1984 Commercial Space Launch Act (Launch Act) and the Land Remote-Sensing Commercialization Act of 1984 (Commercialization Act), the United States space programme has moved to the more recent Artemis Accords.

The Accords are a set of non-binding political commitments, the legal roots of which lay in the 2015 US Commercial Space Launch Competitiveness Act (Tetik, 2023). In sharp contrast to the principles set out in the OST and the Moon Agreement, the law enables United States citizens and companies to own, possess, transport, use and sell resources from outer space, including asteroids, the moon and other celestial bodies. The Act has been enacted to bolster the United States' long-drawn interpretation that neither the OST nor the Moon Agreement explicitly bars outer space mining. Further, as per US lawyers, in the absence of international consensus, countries are free to adopt the interpretation that suits their interests (Neef, 2021).

Two things deserve to be noted at this juncture. The United States' interpretation of the international space law stands on shaky ground considering that principles underlying them like 'the common heritage of mankind' and 'province of the mankind' have been more or less interpreted as a prohibition against private ownership (Arnold, 1975; Lee, 2012). Secondly, no single country can form international law unless supported by other countries. In this respect, it is noteworthy that the United States is not alone in enacting domestic legislation providing for commercial space mining licenses. Other countries like Luxemburg, the UAE and Japan have followed in the footsteps of the United States in enacting similar legislation (DePagter, 2021). However, even an interpretation adopted by four countries cannot override the provisions of an international treaty (OST) having 112 ratifications.

The Artemis Accords form the inherent and crucial part of the 2015 Act. Released by NASA on May 15 2020, the Accords articulate a shared vision to create a safe and transparent environment for exploration, science, and commercial activities for all of humanity to enjoy. Section 1 of the Agreement lays down the areas where research activities, including mining activities, can be carried out in the form of Moon, Mars, Comets, Asteroids and in the transit between these celestial bodies. In this manner, the Agreement limits the area of operation. Further, the agreement lays down broad guidelines as to cooperation, liability, intellectual property and technology transfer. In essence, the Agreement repeats several provisions of the OST and the Moon Agreement by providing for transparency, sharing of knowledge and setting of uniform standards.

However, Section 10 of the agreement stands out as the most contentious and debatable provision. Section 10 lays down that extraction and utilisation of space resources shall be done in accordance with the Outer Space Treaty. For this purpose, the agreement clarifies that the extraction of resources does not amount to national appropriation, and every extraction shall be notified to the Secretary-General of the United Nations and to the public and the international scientific community. In the analysis of the provision, at least two things are notable:

First, the Outer Space Treaty, by virtue of its Articles I and II, prohibits commercial space mining activities (Haywood, 2024; Zell, 2006). The use of the phrase "province of all mankind" forms the crux of the non-appropriation principle being so crucial to the treaty and extending to both public and private entities as well as to celestial bodies and their resources (Zell, 2006). The prohibition has been given a definite shape in the Moon agreement, which explicitly bars private ownership of lunar resources (Blanchette-Seguin, 2017). Given the same, the agreement's reconciliation between resource extraction in outer space and OST appears untenable.

Secondly, the Agreement's interpretation that resource extraction does not amount to national appropriation is inherently flawed. This is on account of two reasons. Firstly, the commercial extraction of natural resources is inherently linked to property rights, whether public or private to make profits. In such a scenario, the creation of property rights by Accords, on one hand, cannot be divorced from national appropriation, on the other hand, which is barred by the Outer Space Treaty. Additionally, the existence of a legal right to property requires sovereign ownership in the first place. In the absence of the latter, the former cannot come into existence (Lefeber, 2016). The US approach to severing private property rights from sovereign ownership is difficult to accept in the face of a lack of any international law providing for space resource ownership (Lefeber, 2016).

Inherent in the wording of the provision is the US approach that the OST does not ban the extraction of space resources. The Commercial Space Launch Competitiveness Act of 2015 is the concrete manifestation of that approach. Suffice it to say that the United States believe in the unilateral interpretation of international law for its use and application (Su, 2017).

The creation of 'safety zones,' under Section 11, follows the same approach. The concept of 'safety zones' was first explained by the Hague Space Resources Governance Working Group 19 in the Building Blocks project in 2019 as a mechanism to avoid harmful interferences or conflicts. They are in the form of demarcated areas between nations and private entities with the purpose of avoiding any conflict as to any research or mining activities. However, in the absence of clear delimitation, nature of activity and duration, safety zones are highly prone to abuse by states and private entities. There is a significant possibility of states and private entities on the basis of the establishment of safety zones (Matusiak, 2023). In other words, safety zones could well be transformed into operational zones and then into territorial zones. They have an inherent tendency to violate the guarantee of non-appropriation given under Article 2 of the OST (Mallowan et al., 2021). Thus, 'safety zones' are precursors to claims of sovereignty by states and private ownership by individuals.

At present, the Accords have been signed by 48 nations, and except for Russia and China, all major space-faring nations have signed the agreement. The agreement represents a significant shift in international space law. In essence, they are symbolic of the evolving framework of soft law in clarifying the issues and filling gaps in the international space regime. With the possibility of more nations joining the agreement, the United States is on the cusp of rewriting space laws. Although the Accords never claimed to lay down new space laws (de Zwart, 2021) or interpret them in a new fashion, their language and interpretation are difficult to reconcile with the existing laws. In particular, the claim that the agreement is rooted in the OST and follows all international laws is a farfetched interpretation. In reality, the Accords sit at the crossroads between global multilateralism and unilateralism (Neef, 2021). The United States, in essence, is trying to juxtapose domestic law over international law by way of unilateral interpretation and propagation (Fidler, 2020).

Thus, the Artemis Accords, to that end, are a tool to garner international support and legitimacy (F. von der Dunk, 2020). Their ultimate purpose is to condone the commercial exploitation of space resources and sovereign rights over parts of celestial bodies. The programme is specifically aimed at pushing the interests of the thriving commercial space industry in the US. There are potential benefits to the US and significant geopolitical implications of the programme. Although, presently, the Accords cannot be accepted as modifying customary international law laid down by the OST, given its high number of ratifications (Mosteshar, 2020), there is every possibility that the Accords may pose a challenge in times to come.

10.4 Ever evolving and Diversifying: India's Space Programme

Soon after independence, India embarked on the journey of transforming society through scientific endeavours and European positivism (Kumar, 2014). Establishing the institutional structure for research in outer space in 1962 through INCOSPAR (Indian National Committee for Space Research), the country built its first rocket launching station, Thumba Equatorial Rocket Launching Station (TERLS) in Trivandrum, Kerala in the same year. The creation of ISRO in 1969 became a turning point in India's space program history. Soon thereafter, the Government of India established the Space Commission and the Department of Space (DOS), wherein the former was made responsible for framing space policies while the latter was tasked with their implementation. At the outset, the country's space programme was initiated and has long continued with an avowed vision of peaceful use of outer space (Chandrashekar, 2011). This has been reiterated by the country by the signing of the Outer Space Treaty, 1967 and at various international meetings. India's approach has been consistent with an avowed object of meeting the country's socio-economic goals through the development of space capabilities (Christensen et al., 2009). However, over the period of time, India's space program has undergone significant changes with the adoption of the space securitisation doctrine and the rise of the 'NewSpace' age (Rajagopalan & Stroikos, 2024).

10.4.1 Militarisation of India's Space Programme

For long, India has followed the policy of non-militarisation of outer space and advocated for peaceful uses of outer space only. However, events like the creation of the Space Force by the US, the test launch of anti-satellite missile systems by China and the proposal to establish a Space Command Centre by France turned India's non-militarist space policy (Rajagopalan, 2011). Accordingly, India established a Defence Space Agency (DSA) and a Defence Space Research Organization (DSRO) in 2019 to coordinate its military space capabilities and to conduct military research, respectively. Proving its defence capabilities, India successfully conducted its first anti-satellite missile test in 2019 through its Mission Shakti. Keeping in line with global developments and evolving threat perception, India is now looking to boost its military space capabilities.

10.4.2 Privatisation of India's Space Sector

The NewSpace Age began in the second decade of the twenty-first century and was characterised by intense competition and contributions by private space companies. Companies like Space X, Blue Origins, Virgin Galactic, etc. in the US have contributed equally in space exploration programmes from private rocket launches to satellite launches. India has not remained immune to this development. Many private players in India have gained the technical competence and experience to deliver equally. However, the commercialisation of space activities in India goes back to 1992, when Antrix Corporation Limited was established as a commercial and marketing arm of ISRO (Nagendra & Basu, 2016). Since then, ISRO has effectively commercialised the country's space launches by successfully launching satellites of various countries and supplying satellite systems and subsystems.

In 2019, NewSpace India Limited (NSIL) was incorporated as a wholly owned Government of India company and as the second commercial arm of ISRO, subsequently replacing Antrix. Promoting private players, a dedicated single window nodal agency, the Indian National Space Promotion and Authorisation Centre (IN-SPACe) was opened to enable greater coordination between ISRO and private entities. The culmination of the privatisation of the space sector was the establishment of the Indian Space Association (ISpA) in 2021 as an advocacy group consisting of major private players. The commercialisation of the space sector and the increased cooperation between the public and private companies are essentially on account of the realisation of enhanced competency and the immense contribution that private companies can make to the Indian space sector.

10.5 India and Artemis Accords: A Union with Deep Differences

India's space industry, just like the others, has not grown in isolation. Time and again, India has cooperated with other countries in providing assistance and in developing its space capabilities. It has partnered with various spacefaring nations and organisations in sustainably using outer space resources to the common benefit of mankind (Guruprasad, 2018). On one hand, India has successfully launched other nations' satellites; on other occasions, it has used powerful rockets from other countries to deliver heavier payloads into outer space. While India has partnered with almost all spacefaring nations, it shares a close bond with the United States, which goes back to the launching of the first sounding rocket by India, Nike-Apache, a rocket developed in the United States (Lele, 2016).

Over time, the India-US space relationship has grown and matured into a strategic partnership in space ventures. Since 2021, both countries have entered into a plethora of agreements to strengthen and develop space capabilities. In this respect, four initiatives deserve special mention: INDUS-X, the Initiative on Critical and

Emerging Technology (iCET), the enhanced NASA-ISRO relationship and the Artemis Accords (Boyd et al., 2024). Signed in September 2023, the Artemis Accords mark a crucial point in the India- US space program relationship. At the same time, the signing of the Accords raises some crucial questions as to India's long-held position on non-appropriation of outer space, its ability to reap the benefits of the agreement and differences in Space programme objectives. These issues, amongst others form part of the present discussion.

10.5.1 Different Space Objectives

Since its inception, the purpose of the Indian space program has been to use outer space to meet the developmental and socioeconomic needs of the country (Rajagopalan & Stroikos, 2024). As such, space projects involving communication, remote sensing and meteorology were given preference over others. In this respect, the main focus has been on three main societal applications of outer space: tele-education, telemedicine and village resource centre programmes (Lele, 2017). With the maturation of the Indian space programme and techniques, deep space endeavours like the Mars probe or the Moon exploration projects were undertaken. Combined with it was the initiation of the commercial utilisation of India's space resources.

A significant development has been the growth and promotion of the private sector in the Indian space industry. In recent years, many private players have immensely contributed to the Indian space growth story. More recently, the securitisation perspective has opened up the avenue for exploring the military uses of outer space, signifying a major shift from the earlier held 'civilian use of outer space' philosophy (Rajagopalan, 2011). In sum, India's space program is aimed at meeting the socioeconomic and security needs of the country, wherein the protection of the environment and lives and the stimulation of public awareness and scientific quest are key ingredients, as apparent from the Indian Space Policy 2023.

On the other hand, the US-designed Artemis Accords, while proclaiming the peaceful and civilian use of outer space, in the end are aimed to promote commercial exploration and exploitation of space resources. The gamut of principles and rules that the agreement lays down as a set of mutually agreed political commitments to facilitate space cooperation is essentially aimed at creating a non-binding global regime for outer space resource exploration and exploitation (Matusiak, 2023). Additionally, the Accords emphasise on the extraction and utilization of resources on comets and asteroids.

In the light of the above discussion, two things are noteworthy. First, the Indian space programme has from its inception and continued to be for the purposes of the socio-economic development of the country (Lele, 2017). Nowhere has the programme spelt out the objective of meeting the needs of the country through outer space resource mining. Secondly, over the years, the country's space programme has matured to diversify its purpose to include scientific quests on celestial bodies like the Moon and Mars (Baluragi & Suresh, 2020). To what extent, the programme is capable of exploring comets or asteroids is a big question to answer. Taken together, the objectives of the agreement do not match with that of the Indian space programme and hence raise doubts over the efficacy of the decision to sign the agreement.

10.5.2 Less Developed Private Space Industry

In the last two decades or so, the private sector has significantly stepped up its role in the country's space programme. Thanks to the liberalisation and privatisation policy of major spacefaring nations, private space companies have seen phenomenal growth in recent years. Whether it is India or the United States, the importance of private space firms in the development of the country's space programme has been established beyond doubt (Iacomino, 2019). Despite the recognised importance, the role and capabilities of the private space companies significantly differ in the two countries.

Beginning in the 1980s, the United States' private space sector has become a behemoth over a period of time. Its enormous size makes up 80% of the US space industry. Companies like SpaceX have mastered significant technological competencies in reducing orbital costs, developing reusable rockets and running manned space missions (Weinzierl et al., 2020). So much so that the country's space programme is now dependent on privately developed rockets and their missions to meet its requirements. The development of rocket technology in the form of Falcon Heavy and Dragon has been instrumental in pivoting SpaceX as a giant in the space industry. However, SpaceX is not the only mega private space player. For a long time, the US space arena has been dominated by two big names: Boeing and Lockheed Martin wherein they have successfully contributed to the needs and requirements of NASA at various levels. Thus, the United States private space sector is quite well developed and mature with equally capable technologies and competencies (Chaddha, 2010).

On the other hand, the Indian private space sector has seen growth only in the past decade or so, with some relaxation of government regulations in 2010. Until then, the private sector has acted only as Tier-2 or Tier-3 suppliers (Boyd et al., 2024). The year 2019 marked a decisive year in the area of the promotion of the private sector with the establishment of New Space India Limited (NSIL) and the Indian National Space Promotion and Authorisation Centre (IN-SPACe). These efforts have borne fruit in the form of the launch of India's first privately developed rocket, Vikram-S, in 2022 (Rajagopalan, 2022). A more nuanced approach towards the development of India's private space industry was taken with the release of the

Indian Space Policy 2023. However, with the majority of Indian space ventures being small start-ups, their capabilities are severely limited.

As is evident from the above discussion, the Indian private space sector is still in its nascent stage (Nagendra & Basu, 2016). It significantly lacks the technological and experiential capabilities to run space missions on its own. Unlike its US counterpart, the private sector in India is heavily dependent on the support and guidance of the public sector. It is noteworthy that the Indian private space industry, being a capital-intensive industry, is largely dependent on foreign investments for its financing and growth (Boyd et al., 2024). In light of such glaring inadequacies, it is questionable to what extent India would be able to reap benefits out of the Artemis Accords, given that the Accords are primarily aimed at the commercialisation of outer space activities.

10.5.3 Adherence to the Principle of Non-Appropriation

India is a party to the 1967 Outer Space Treaty and has time and again reiterated its steadfast support for the principle of non-appropriation and denial of sovereign rights over celestial bodies. Against such a stand, the provision of 'safety zones' outlined in the Artemis Accords becomes problematic for India. At the outset, although the Accords promise to adhere to the principle of non-appropriation and other international laws, the ambiguity surrounding the creation of 'safety zones' has the inherent tendency to challenge the principle of non-appropriation (Gilbert, 2023). Defined as areas of nominal operations to avoid harmful interference and resultant conflicts, 'safety zones' resemble colonial agreements amongst colonists to carve out colonies for themselves. At the very least, if not ownership, 'safety zones' grant possession rights over celestial bodies (O'Brien, 2020).

That safety zones are the precursor to claiming ownership rights or appropriation rights over celestial bodies is evident from the objectives of the United States' private space industry itself. Take the case of SpaceX, which has not hidden its intention to colonise Mars (Cuthbertson, 2020). Such attempts of colonisation and claims of sovereign rights run absolutely contrary to international law (Salmeri, 2020). Further, the notion of exploitation and, thereby private ownership of space resources is itself questionable. The modern right to private property is intricately linked to sovereignty and, therefore any grant of private property without the concomitant sovereign right is dubious in the least and illegal at the most (Przeworski & Wallerstein, 1986).

Read together, the concept of 'safety zones' and 'private ownership' of space resources run contrary to India's long-standing position of non-appropriation and ownership rights over celestial bodies. The signing of the Accords by India to that end raises questions on India's policy.

10.5.4 No Domestic Space Law

Article VI of the Outer Space Treaty (OST) imposes national responsibility for activities in outer space. As such, in order to regulate national activities in outer space and bilateral commercial ventures, many spacefaring nations have designed and adopted dedicated space policies. The United States, for instance, passed the U.S. Commercial Space Launch Competitiveness Act in 2015. The Act specifically authorises and regulates commercial activities by private entities in outer space. Additionally, the United States has passed a number of policies to supervise and promote the growth of its space sector. On the contrary, India still does not have a national space law. There is no dedicated law in India providing for the exploration and use of outer space (Mund, 2021).

Space regulation in India has generally been an ad hoc mechanism (Kaul & Jakhu, 2010). The regulation has largely been in terms of space policies like the Satellite Communication Policy and the Remote Sensing Data Policy. While the former has aimed at the development of a satellite industry in the country, the latter creates a licensing regime with regard to remote sensing satellites and data. Besides, the country has some odd 74 space laws that govern different aspects of space activities, including by private players. This vast body of ad hoc space regulatory mechanisms has been proven to be ineffective in the all-around development of India's space sector (Athar ud din, 2021). There is an urgent need for a national space law that can serve as a guiding mechanism for India's space objectives, inter-state cooperation and private sector participation.

In this regard, the Draft Space Activities Bill, 2017, was introduced. The bill consists of crucial provisions for the development of the private sector, national and international obligations, liability framework, licensing regime and safety mechanisms. However, the bill is still awaiting cabinet clearance and thereafter, parliamentary approval. Even if the bill is passed in its current form, it fails to address critical questions pertaining to possession rights by private players (Athar ud din, 2021). In the absence of such provisions, the bill is inadequate in comprehensively regulating the activities of private players, particularly in light of agreements like the Artemis Accords that primarily deal with commercial space exploration by the private sector.

10.5.5 Abandoning of the Non-Alignment Doctrine

India's space adventure began amidst the intense Cold War between the US and the USSR. Surprisingly enough, not only both the superpowers but also France helped India find its feet in the form of the establishment of the country's first space facility – Thumba Equatorial Rocket Launching Station (TERLS). Time and again, India has successfully cooperated with various spacefaring nations. Barring China, India has partnered with one or the other nations in space exploration at some point in the past.

In this regard, what is noteworthy was the effective use of its non-alignment policy that not only enabled India to carry out its cooperative endeavours but also pushed through the United Nations bureaucratic hurdles in getting the necessary sponsorship (Siddiqi, 2015). Over the years, the country's space capabilities have grown to enable India to provide crucial launch and technological support to other countries. In this manner, the policy of non-alignment has benefitted India for a long period of time (Pant & Super, 2019). Thus, India's space programme is firmly rooted in its non-alignment policy.

In this regard, India's decision to sign the Artemis Accords and thereby cooperate with the United States in commercial space resource exploration and exploitation runs counter to its long-standing policy of non-alignment. This is more so in view of the fact that Russia along with China has launched the International Lunar Research Station (ILRS) initiative in response to the Artemis Accord. The ILRS aims to create a sustainable lunar base on the moon in the form of a research facility on the moon, around its orbit and on the Earth. The purpose of the initiative is to conduct scientific research, observations and experiments about the composition and resources of the moon. Similar to the Artemis Accords, ILRS is open to all countries to join through negotiation with Russia and China.

10.6 Conclusion

Suffice it to say that the Artemis Accords mark a dramatic turn in the development of international space law. The Accords, in many ways, have the tendency to disrupt international space regimes and space collaborations. The Accords pave the way for unilateral interpretation and application of international law, a precedent dangerous to international comity and space cooperation. Secondly, the Accords, whether intentionally or unintentionally, widened the east-west rift. Against such a background, the signing of the Accords by India marks a defining moment in India's space history. The step especially puts India in a difficult position whereby on account of its technological inadequacy, it may not be able to reap any benefits from the agreement but may end up losing important space allies. There is no denying the fact that privatisation and commercialisation are necessary tools for progress and development in today's world. However, such progress cannot come at the cost of defying international law and choosing one nation over the other. To such an extent, any decision by India must be balanced against its long-term goals and space partnership, therefore calling for greater prudence and understanding.

References

Akpambang, E. M., & Akanle, A. O. (2023). Outer space commercial mining: Ownership rights and contemporary legal issues. *Scholars International Journal of Law, Crime, and Justice, 6*, 343.

Arnold, R. P. (1975). The common heritage of mankind as a legal concept. *Journal of Space Law, 9*, 153.

Athar ud din. (2021). India's Quest for a National Space Law and the Missing Piece of Possessory Rights. *India Quarterly, 77*(4), 642–660.

Baluragi, H., & Suresh, B. N. (2020). Indian space program: Evolution, dimensions, and initiatives. In Handbook of Space Security (pp. 1–19). Springer.

Barnes, R. (2009). *Property rights and natural resources*. Hart Publishing.

Blanchette-Seguin, V. (2017). Reaching for the moon: Mining in outer space. *New York University Journal of International Law and Politics, 49*.

Boyd, J., et al. (2024). The future of U.S.-India private sector space collaboration. *IPPR, 5*, 19.

Brünner, C., & Soucek, A. (2012). Outer space in society, politics and law (Vol. 8). Springer Science & Business Media.

Chaddha, S. (2010). US commercial space sector: Matured and successful. *Journal of Space Law, 36*, 19.

Chandrashekar, S. (2011). India and the peaceful uses of outer space. *India Review, 10*, 440.

Christensen, I. A., Hay, J. W., & Peura, A. D. (2009). National development through space: India as a model. In *Space technologies for the benefit of human society and Earth* (p. 453).

Coffeyt, S. (2009). Establishing a legal framework for property rights to natural resources in outer space. *Case Western Reserve Journal of International Law, 41*.

Connolly, R., & Bennett, A. (2022). The next arms race and the unknown frontier of outer space: The conceptual challenges for international law and space weaponization. *Journal of Space Law, 46*, 294.

Cuthbertson, A. (2020, February 28). Elon Musk's SpaceX will 'make its own laws on Mars' | The Independent. Independent. https://www.independent.co.uk/space/elon-musk-spacex-mars-laws-starlink-b1396023.html

De Man, P. (2017). State practice, domestic legislation, and the interpretation of fundamental principles of international space law. *Space Policy, 42*, 92.

DePagter, M. M. (2021). "Who Dares, Wins:" How Property Rights in Space Could Be Dictated by the Countries Willing to Make the First Move. Chi. J. Int'l L. Online, 1, 116.

De Zwart, M. (2021). To the moon and beyond: The Artemis Accords and the evolution of space law. In *Commercial and military uses of outer space* (p. 65).

De Zwart, M., Henderson, S., & Neumann, M. (2023). Space resource activities and the evolution of international space law. *Acta Astronautica, 211*, 155.

Deplano, R. (2021). The Artemis Accords: Evolution or revolution in international space law? *International & Comparative Law Quarterly, 70*, 799.

Fidler, D. P. (2020). *The Artemis Accords and the next generation of outer space governance*. Council on Foreign Relations.

Feichtner, I. (2019). Mining for humanity in the deep sea and outer space: The role of small states and international law in the extraterritorial expansion of extraction. *Leiden Journal of International Law, 32*, 255.

Gabrynowicz, J. I. (2010). One half-century and counting: The evolution of US national space law and three long-term emerging issues. *Harvard Law & Policy Review, 4*, 405.

Gilbert, A. Q. (2023). Implementing safety zones for lunar activities under the Artemis Accords. *Journal of Space Safety Engineering, 10*, 103.

Guruprasad, B. R. (2018). Understanding India's international space cooperation endeavour: Evolution, challenges, and accomplishments. *India Quarterly, 74*, 455.

Hasin, G. (2022). From 'space law' to 'space governance': A policy-oriented perspective on international law and outer space activities. *SSRN Journal.*. https://www.ssrn.com/abstract=4179934

Haywood, J. (2024). Ocean: A History of the Atlantic Before Columbus. Bloomsbury Publishing.

Hobe, S., & De Man, P. (n.d.). *National appropriation of outer space and state jurisdiction to regulate the exploitation, exploration, and utilization of space resources*.

Holton, G. J., & Brush, S. G. (2001). Physics, the human adventure: From Copernicus to Einstein and beyond. Rutgers University Press.

Iacomino, C. (2019). The evolving role of private actors in space exploration. In *Commercial space exploration: Potential contributions of private actors to space exploration programmes* (p. 35).

Kaul, R., & Jakhu, R. S. (2010). Regulation of space activities in India. In National Regulation of Space Activities (pp. 153–198). Springer.
Kostenko, I. (2021). Artemis Accords and the future of space governance: Intentions and reality. *Advanced Space Law, 8.*
Klare, M. T. (2012). *The race for what's left: The global scramble for the world's last resources.* Metropolitan Books.
Kumar, G. (2014). India's Space Programmes: Remapping Contours of 'Security.' South Asian Journal of Diplomacy, 2229–3361
Lee, R. (2012). Law and regulation of commercial mining of minerals in outer space (Vol. 7). Springer Science & Business Media.
Lele, A. (2016). Power dynamics of India's space program. *Astropolitics, 14*, 120.
Lele, A. (2017). India's policy for outer space. *Space Policy, 39*, 26–32.
Lefeber, R. (2016). Relaunching the Moon Agreement. *Air and Space Law, 41.*
Mallowan, L., Rapp, L., & Topka, M. (2021). Reinventing treaty-compliant 'safety zones' in the context of space sustainability. *Journal of Space Safety Engineering, 8*, 155.
Masson-Zwaan, T. L., & Hobe, S. (2010). The Law of Outer Space: An Experience in Contemporary Law-Making, by Manfred Lachs, Reissued on the occasion of the 50th anniversary of the International Institute of Space Law. Brill.
Matusiak, M. (2023). Artemis Accords–a new era of space law or an initiative of the world's superpower? *Studia Iuridica, 97*, 168–184.
Meadows, D., & Randers, J. (2012). The limits to growth: The 30-year update. Routledge.
Mosteshar, S. (2020). Artemis: The discordant Accords. *Global Sustainability, 44.*
Mund, P. (2021). 'Tomorrow is today' for the Indian space saga: Delineating the legal framework for space activities in India. *Air and Space Law, 46.*
Nagendra, N. P., & Basu, P. (2016). Demystifying space business in India and issues for the development of a globally competitive private space industry. *Space Policy, 36*, 1.
Neef, R. (2021). Artemis Accords: A New Path Forward for Space Lawmaking? *Adelaide Law Review, 42*(2).
Nelson, T. G. (2010). The Moon Agreement and private enterprise: Lessons from investment law. *ILSA Journal of International & Comparative Law, 17*, 393.
O'Brien, D. (2020). The Artemis Accords: Repeating the mistakes of the age of exploration. *The Space Review* (Online).
Osada, Y. (2022). Governance of space resources activities: In the wake of the Artemis Accords. *Georgetown Journal of International Law, 53.*
Pant, H. V., & Super, J. M. (2019). Non-alignment and beyond. New Directions in India's Foreign Policy: Theory and Praxis, 127–148.
Pascale, M. (2023). Space, the Final Frontier: Navigating the Complexities of Commercial Spaceflight, Resource Extraction, and Colonization. U. Ill. JL Tech. & Pol'y, 151.
Prema, E., Bharathi, R. V., & Sundar, V. S. (2024). The Artemis Accords: A New Phase in the Development of Outer Space Law. In Ascending to Space: Critical Perspectives from New Zealand and other Nations (pp. 73–97). Springer.
Przeworski, A., & Wallerstein, M. (1986). Popular sovereignty, state autonomy, and private property. *European Journal of Sociology, 27*, 215.
Qizhi, H. (1997). The outer space treaty in perspective. J. Space l., 25, 93.
Rajagopalan, R. P. (2011). India's changing policy on space militarization: The impact of China's ASAT test. *India Review, 10*, 354.
Rajagopalan, R. P. (2022, November 21). Skyroot creates a new 'Prarambh' for Indian space. Orfonline.Org. https://www.orfonline.org/expert-speak/skyroot-creates-a-new-prarambh-for-indian-space
Rajagopalan, R. P., & Stroikos, D. (2024). The transformation of India's space policy: From space for development to the pursuit of security and prestige. *Space Policy, 69*, 101633.
Siddiqi, A. (2015). Making space for the nation: Satellite television, Indian scientific elites, and the Cold War. *Comparative Studies of South Asia, Africa and the Middle East, 35*(1), 35–49.

Raclin, G. C. (1986). *From ice to ether: The adoption of a regime to govern resource exploitation in outer space*. International Law.

Salmeri, A. (2020, December 5). No, Mars is not a free planet, no matter what SpaceX says. *SpaceNews*. https://spacenews.com/op-ed-no-mars-is-not-a-free-planet-no-matter-what-spacex-says/

Su, J. (2017). Legality of unilateral exploitation of space resources under international law. *International & Comparative Law Quarterly, 66*, 991.

Tetik, B. (2023). The Outer Space as a Domain of Competition and Cooperation From the Cold War to Today [Master's]. Middle East Technical University.

Von der Dunk, F. G. (2020). Scoping national space law: The true meaning of "National activities in outer space" of Article VI of the Outer Space Treaty.

Von der Dunk, F. G. (2023). Property rights over the Moon or on the Moon? The legality of space resource exploitation on celestial bodies. *JL & Innovation, 6*, 95.

Weinzierl, M. C., Lucas, K., & Sarang, M. (2020). SpaceX, economies of scale, and a revolution in space access. Harvard Business School Case, 720–027.

Winchester, S. (2021). *Land: How the hunger for ownership shaped the modern world*. William Collins.

Zell, J. L. (2006). Putting a mine on the Moon: Creating an international authority to regulate mining rights in outer space. *Minnesota Journal of International Law, 15*, 489.

Chapter 11
Celestial Property Rights in Outer Space: A TWAIL Exploration

Aaditya Vikram Sharma

Abstract The evolving dynamics of international space law, especially in the context of celestial property rights, demand a nuanced analysis through the lens of Third World Approaches to International Law (TWAIL). This study embarks on an explorative journey to understand how TWAIL can provide a critical perspective on the principles and practices governing the outer space regime, particularly focusing on the issue of property rights on celestial bodies. At the heart of international space law are the 1967 Outer Space Treaty and subsequent agreements, which collectively advocate for the non-appropriation of outer space by any one country, emphasizing its use for the benefit of all humankind. However, the rapid advancements in space technology and the increasing participation of private actors and countries in space exploration raise intricate questions about the applicability and adequacy of these legal frameworks in addressing contemporary challenges, including the equitable distribution of outer space resources. By employing the TWAIL framework, this study critiques the existing international space law from the perspective of countries in the Global South, arguing that the current legal norms reflect the interests and viewpoints of the developed world, potentially marginalizing the aspirations and concerns of developing countries. It scrutinizes the concept of celestial property rights against the backdrop of colonial histories and the principles of sovereignty and self-determination, highlighting the parallels between terrestrial and celestial resource exploitation. Furthermore, this analysis delves into the discourse on the common heritage of mankind, equity, and environmental considerations in space activities, proposing a TWAIL-informed approach to international space law that promotes a more inclusive, equitable, and sustainable governance of outer space resources. This study ultimately calls for a reimagined legal framework that aligns with the principles of justice, equality, and international cooperation, ensuring that the benefits of space exploration are shared equitably among all nations.

A. V. Sharma (✉)
Vivekananda Institute of Professional Studies – Technical Campus, New Delhi, India
e-mail: aaditya.sharma@vips.edu

S. Bhat B., A. Variath (eds.), *Artemis Accords and Resource Mining in Outer Space*, Issues in Space, https://doi.org/10.1007/978-981-95-0152-6_11

Keywords TWAIL · Artemis accords · Celestial property rights · Outer space treaty · Equitable resource distribution · Inequalities

11.1 Introduction

The Third World Approaches to International Law (TWAIL) is a perspective that critically examines the role and impact of international law from the viewpoint of the world's marginalised and less economically developed countries (Anghie, 2023). TWAIL posits that the current global legal order perpetuates colonial legacies and systemic inequalities, often privileging the interests of dominant states while sidelining those of the Global South (Appiagyei-Atua, 2015). It advocates for reimagining international law to ensure it promotes justice, equality, and development for all countries (Anghie, 2023). Central to this approach is the recognition of historical injustices and the imperative to address imbalances in global power and resource distribution (Enyew, 2022). TWAIL scholars advocate for a paradigm shift in international law, suggesting that it must be reoriented to genuinely reflect and respond to aspirations of the Global South (Modirzadeh, 2023; Anghie, 2023). This involves not only critiquing the ways international law has reinforced global disparities but also actively promoting legal standards that champion sustainable development, human rights, and social justice (Chimni, The international law of jurisdiction: A TWAIL perspective, 2022; Fidler, 2003). Such a reimagined legal framework would require a significant shift in how international law is conceptualised, taught, and applied, ensuring it becomes a tool for genuine transformation and empowerment of marginalised nations (Chimni, 2003).

The development of International Space Law (ISL) and its interaction with TWAIL represents a fascinating area of legal scholarship. As space law evolves to address new challenges, such as space tourism, space mining, and a permanent presence of human beings on celestial bodies, the TWAIL approach can emphasise the importance of ensuring that these developments do not perpetuate global inequalities (Vazhapully, 2024). Herein, TWAIL-ers argue for the inclusion of third-world perspectives in the creation and application of space law, ensuring equitable access to outer space resources and technology while preventing the dominance of space by a few powerful nations. This approach seeks to guarantee that the benefits of space exploration and utilisation are shared by all humanity, reflecting a commitment to justice and international cooperation in the final frontier.

This chapter begins by exploring the foundational principles of ISL, setting the stage for a critical analysis through the lens of Third World Approaches to International Law (TWAIL). It introduces the TWAIL framework, highlighting its relevance in evaluating space law from the perspective of developing nations. Building on this conceptual groundwork, the chapter adopts qualitative and analytical methodologies to investigate the contributions and positions of Global South countries during the evolution of international space law. The primary objective is to discern whether a distinct "Third World" perspective emerges by examining these states' roles, viewpoints, and negotiating strategies in the drafting of space treaties. Central to this inquiry is the contested issue of celestial property rights, a topic that has gained urgency amid the rapid expansion of space exploration and commercialization (Tronchetti, 2009; Su, 2022). To address this, the chapter critically engages with the two key treaties that govern celestial property rights: the 1967 Outer Space Treaty and the 1979 Moon Agreement, analyzing how these agreements reflect the equitable distribution of space resources and the broader aspirations of developing nations.

11.2 Framework and Approaches

11.2.1 International Space Law

The history of ISL began in the mid-twentieth century as space exploration became technically feasible, leading to the need for a legal framework to govern activities beyond Earth's atmosphere (Jain et al., 2022). The United Nations (UN) played a pivotal role in establishing the Committee on the Peaceful Uses of Outer Space (COPUOS) in 1959 (UN, COPUOS history, 2024). The key treaties of ISL are the Treaty on Principles Governing the Activities of States in the Exploration and Use of Outer Space, Including the Moon and Other Celestial Bodies, 1967 (OST), the Agreement on the Rescue of Astronauts, the Return of Astronauts and the Return of Objects launched into Outer Space, 1968 (Rescue Agreement), Convention on International Liability for Damage Caused by Space Objects, 1972 (Liability Convention), Convention on Registration of Objects Launched into Outer Space, 1975 (Registration Convention), and the Agreement Governing the Activities of States on the Moon and Other Celestial Bodies Moon Agreement, 1979 (Moon Agreement). Taken together, these five treaties form the foundation of ISL, setting principles for the exploration and use of outer space, emphasising that it should be carried out for the benefit of all countries and prohibiting the placement of nuclear weapons in space (UNOOSA, International Space Law: United Nations Instruments, 2017). These legal instruments reflect the international community's commitment to peaceful cooperation in space exploration, yet challenges remain in addressing new issues like space debris, commercial space activities, and resource exploitation (Jakhu & Pelton, Global Space Governance: An International Study, 2017). For the purposes of this study, only the 1967 OST and the 1979 Moon Agreement are considered, as other ISL treaties do not address issues related to celestial property rights.

The OST was adopted in 1967. It establishes the framework for ISL, prohibiting the placement of nuclear weapons in space, limiting the use of the Moon and other celestial bodies for peaceful purposes, and declaring that outer space should be accessible to all countries. The OST emphasises that outer space is the province of mankind, prohibiting any country from claiming sovereignty over outer space or any celestial body (UNOOSA, 2022). On the other hand, the Moon Agreement was adopted in 1979 to extend the provisions of the OST specifically to the Moon and other celestial bodies within the solar system, excluding Earth. It emphasises that the Moon and such celestial bodies should be used exclusively for peaceful purposes, prohibits any state from claiming sovereignty, and outlines that their exploration and use should benefit all countries (UNOOSA, 2024). Article 11 of the Agreement also introduces the concept of the Moon and its natural resources as the Common Heritage of Mankind (CHM), suggesting that an international regime should be established to govern the exploitation of such resources. However, unlike the OST, it has been ratified by only seventeen nations, reflecting broader debates about its provisions and practical implementation (COPUOS, Status of International Agreements relating to activities in outer space as at 1 January 2024, 2024). Beyond the foundational treaties and agreements, various supplementary instruments and frameworks also guide the governance of space activities. These include numerous guidelines, principles, and resolutions adopted by the UN General Assembly.

As the frontier of space exploration expands, ISL faces emerging challenges, particularly in adapting to the rapid advancements in space technology and the increasing involvement of private entities in space activities. The principles established in the foundational treaties must now address issues such as the commercial exploitation of space resources, orbital space debris management, regulation of private spaceflight, and celestial property rights. These developments necessitate a dynamic and forward-looking approach to space law, ensuring that it not only promotes the peaceful use of outer space but also fosters responsible and sustainable activities in the space environment. At the same time, it is necessary to look into the past and see the views of all nations so that the idea of 'space law' can be re-addressed.

11.2.2 The Third World Approaches to International Law

The Third World Approaches challenge the notion of universalism in international law, which often serves as a vehicle for imposing Western values and norms on non-Western societies. For TWAIL-ers, this supposed universality is a guise for maintaining Western hegemony in international legal discourse. Scholars advocate for a pluralistic approach that recognises the diversity of legal and cultural traditions in the Third World, arguing that a one-size-fits-all model is both inappropriate and unjust (Pahuja, 2011; Kapur, 2023). Further, one of the fundamental aspects of TWAIL is its critique of the colonial legacies embedded within international law. TWAIL scholars argue that many contemporary legal norms and institutions were shaped during the colonial era and continue to reflect colonial power dynamics (Chimni, 2003). Scholars like B.S. Chimni and James Gathii emphasise the importance of decolonising legal thought, highlighting how colonialism has shaped the legal and political realities of the Global South (Gathii, War, Commerce, and International Law, 2009). TWAIL strongly emphasises praxis, encouraging scholars to engage actively in social and political struggles. This involves moving beyond theoretical critiques to practical efforts to reshape international law to serve the interests of Third World countries better. TWAIL scholars see themselves as thinkers and activists committed to transformative justice and the empowerment of marginalised communities (Natarajan et al., 2018). Richard Falk and Mohsen al Attar are notable proponents of this activist orientation within TWAIL (Falk, 2016; al Attar, 2021).

TWAIL is said to have begun after the Bandung Conference of 1955. By the late twentieth century, it emerged as a leading critique from scholars in Asia, Africa, and Latin America who sought to transform existing international legal frameworks that primarily reflected Western interests (Obiora Chinedu, 2008). A pivotal moment in the development of TWAIL was the 1995 Cairo conference, which helped shape the movement's agenda and fostered collaboration among scholars from the Global South (Falk, 2016). Influential figures such as Antony Anghie, Makau Mutua, Upendra Baxi, R.P. Anand, B.S. Chimni, and Y.K. Tyagi, to name a few, have been instrumental in developing TWAIL's critical perspective, emphasising the need to decolonise international law (Mutua, 2000).

TWAIL operates as both a theory and a methodology, offering a framework for analysing international law through the lens of historical injustices and power dynamics. It draws from various intellectual traditions, including post-colonial theory, critical race theory, and historical materialism, to critique the structures of international law. By utilising empirical evidence and logical reasoning, TWAIL provides a predictive and testable model that seeks to reimagine international law from a Third World perspective (Okafor, 2008). One of the key methodologies employed by TWAIL scholars is historical and empirical analysis. This approach involves examining the historical context of international legal norms and institutions, highlighting how they have been shaped by colonialism and imperialism. By critically analysing historical events and their impact on contemporary legal frameworks, TWAIL scholars aim to uncover the biases and inequities inherent in international law (Anghie, 2007). Scholars like Obiora Okafor and James Gathii have contributed significantly to discussions on global commons and resource equity (Gathii, 2020).

However, while TWAIL has been influential in reshaping international legal discourse, it is not without criticism. Some argue that its focus on deconstruction may at times lack concrete proposals for reconstruction. In contrast, others note that the broad scope of TWAIL may dilute its effectiveness, as it encompasses diverse methodologies and perspectives (Al Attar, 2020; Haskell, 2014; Burra, 2016). However, the critiques also highlight the movement's strength in fostering a plurality of voices and approaches within international law (Emtseva, 2022).

11.2.3 Interplay between Third World Approaches and Space Law

The intersection of TWAIL and ISL remains a relatively nascent area of scholarship. While extensive research has been conducted applying TWAIL principles to fields such as human rights law, trade law, international tribunals, the UN system, and property rights, particularly through frameworks like Trade-Related Aspects of Intellectual Property Rights (TRIPS), international financial institutions, the transnational corporate sector, redistribution of property rights, and the evolution of the CHM principle (Chimni, 2003), the application of TWAIL to space law is comparatively underexplored. However, it would be incorrect to suggest that this area is entirely overlooked. Scholars such as Sandeepa Bhat, Cassandra Steer, P. A. Adar, Timiebi Aganaba-Jeanty, Cristian van Eijk, Cait Storr, Arthur Morland, and Kiran Mohan Vazapully have contributed significantly to bridging this gap by employing the TWAIL methodology to analyse and critique existing space law frameworks (Steer, 2019; Bhat, 2009; Aganaba-Jeanty, 2016; van Eijk & Aganaba, 2021; Storr, 2021; Morland, 2024; Vazhapully, 2024).

Further, Prof. B.S. Chimni in his seminal work titled the "Third World Approaches to International Law: A Manifesto", refers to the global commons wherein he highlights the challenges being faced by the third world (Chimni, 2003). Of course, this concept can only be applied on a tangent as it is unclear whether outer space legally represents a global common (Pic et al., 2023). Antony Anghie's comments in his foreword in the European Journal of International Law challenge this interpretation. Anghie points out that outer space and the global commons are usually seen from the lens of property rights (Anghie, 2023). Taking this idea forward, TWAIL's application to ISL can directly be made with reference to celestial property rights.

11.3 Development of Space Law and the Third World

The evolution of ISL closely paralleled the Space Race which was a period marked by intense competition between the United States (US) and the Union of Soviet Socialist Republics (USSR) to achieve significant milestones in space exploration. This era began with the USSR's launch of Sputnik I in 1957, the first artificial satellite, which intensified global concerns about the potential militarisation of space. The international community, apprehensive about extending Cold War rivalries into outer space, recognised the necessity for a legal framework to ensure the peaceful use of this new frontier.

In response, the UN took proactive measures to address these concerns. In 1959, it established the COPUOS to promote international cooperation and develop space law. This initiative led to the formulation of several key treaties, most notably the OST. This treaty set foundational principles, such as prohibiting the placement of nuclear weapons in space and affirming that the exploration of outer space should benefit all countries. As mentioned in the Introduction, this was followed by four more treaties- the last of which was the Moon Agreement.

This chapter will explore the connection between TWAIL and ISL through empirical analysis. While one can adopt a theoretical lens, that is, following the approach of the previously cited scholars, this chapter instead examines nations' actions and legal position(s). To that end, the following section summarises the stances of developing countries during the negotiation of relevant space treaties.

Since this research focuses on celestial property rights, the analysis is confined to the OST and the Moon Agreement. Within this scope, priority is given to the perspectives of developing countries on property rights in outer space, although, where necessary, the submissions and viewpoints of other nations have also been incorporated. Further, third world countries shall be the ones that were (or are) part of the Non-Aligned Movement (NAM).

11.3.1 The Ad Hoc COPUOS and the Third World

On 13 December 1958, a report was presented by Mr. Matsch, an Austrian national and the Rapporteur of the First Committee, to the UN General Assembly (UNGA). Subsequently, the UNGA adopted Resolution 1348 (XIII) on the Question of the peaceful use of outer space (UNGA, Res. 1384 (XIII) on the Question of the peaceful use of outer space, 1958a). The resolution established an eighteen-member *Ad Hoc* Committee on the Peaceful Uses of Outer Space (COPUOS). However, the USSR decided to boycott the grouping on account of its American-centric membership (Cheng, 1997).

Three third-world countries were part of the COPUOS, i.e., India, Iran and the United Arab Republic (UAR). However, India and the UAR did not attend the meetings, as in their opinion, without the presence of the USSR, no helpful work would have occurred in the Committee's work (Lyall & Larsen, 2009). It was opined that the two states may have stayed away so as not to get involved in a cold war squabble (Jessup & Taubenfeld, 1959). This view seems correct as non-alignment was already gaining traction since the Afro-Asian Conference of 1955 (Dinkel, 2018). In fact, as early as the first session of the COPUOS in 1958, Burma, India and UAR had tabled their own agreement on a consensus method to break the deadlock between the two superpowers (UNGA, 1958b). Ghana and Indonesia supported their view. Thus, even in this early stage of the development of ISL, by not joining the proceedings of the *Ad hoc* Committee, the third-world countries were emphasising a holistic approach. This was not about the substance of the probable international agreement but rather the method of its creation. In other words, a genuinely acceptable agreement would only be possible if both the space powers were on board. This idea reflects a pragmatic and utilitarian approach towards international law building. It was also a breath of fresh air in the Cold War rivalry that dominated at that time.

Notwithstanding, on 14 July 1959, the *Ad hoc* Committee released its report with the contributions of the rest of the members. While the complete analysis of the report is not warranted in this chapter, the Committee did recommend that "serious problems could arise if States claimed, on one ground or another, exclusive rights over all or part of a celestial body.. some form of international administration over celestial bodies might be adopted" (UNGA, 1959a). India appreciated the work done by the Committee but also mentioned that the result(s) would have been better with the participation of both the US *and* USSR (UNGA, Comments on the Report of the Ad Hoc Committee on the Peaceful Uses of Outer Space, 1959b).

As a consequence of the report, the UNGA established a permanent COPUOS that now comprised twenty-four members. These members included developing countries such as Argentina, Brazil, India, Iran, Lebanon, Mexico and the UAR. Among other things, the Committee was tasked to "study the nature of legal problems which may arise from the exploration of outer space" (UNGA, 1959c).

11.4 The Outer Space Treaty

Once the permanent UNCOPUOS was created, work immediately began on an ISL treaty. Both the Americans and the Soviets were majorly interested as they were pioneering space exploration at the time. For instance, on 12 September 1962, during a speech at Rice University, the then-US president, Mr. John F. Kennedy, also announced that there would be an American on the Moon by the end of the decade (Kennedy, 1962).

So, the year 1962 heralded the beginning of the work on the OST. In the COPUOS meetings, representatives from countries such as India, Brazil, and the UAR emphasised that space exploration should benefit all humanity, irrespective of economic development levels. They advocated for non-aggression, non-interference, and equal rights in space activities, proposing international regulations to ensure peaceful uses and prevent militarisation. India highlighted the need for legal frameworks that promote peaceful coexistence and sovereignty while prohibiting propaganda and espionage in space. Brazil supported the establishment of international sounding rocket facilities, emphasising collaboration with more developed countries for technical and financial assistance. The UAR introduced a draft code to guide peaceful cooperation, underscoring the importance of space law in fostering global harmony (UNGA, 1962a). Developing nations stressed the urgency of addressing liability for space vehicle accidents and the rescue of astronauts, recognising these as crucial components of space law (COPUOS, 1962). They also called for increased participation of developing countries in space activities, suggesting that the UN and specialised agencies play a central role in training and providing technical assistance (UNGA, Report of the Committee on the Peaceful Uses of Outer Space, 1962a).

During the meetings of the First Committee, representatives from developing countries played a crucial role by emphasising the importance of international cooperation and the peaceful use of outer space. Nations such as Indonesia and Sudan advocated for global collaboration in space exploration, underscoring that it should not be monopolised by superpowers but should include all countries, regardless of their development status (UNGA, UN Doc A/C.1/SR. 1297, 1962b). This perspective was aligned with UN Resolution 1721 (XVI), which highlighted that space activities should benefit humanity as a whole. Additionally, delegates from countries like Pakistan and Brazil repeated the call for the establishment of international sounding rocket facilities and emphasised the need for training and capacity building for technical workers from underdeveloped regions (UN, Official records of the General Assembly, Seventeenth Session, First Committee, 1290th Meeting, 1962a).

These countries also expressed concerns about the misuse of satellites for propaganda, advocating for regulation under UN supervision to prevent the dissemination of harmful content (UN, Official records of the General Assembly, Seventeenth Session, First Committee, 1295th Meeting United Nations, 1962. Official records of the General Assembly, Seventeenth Session, First Committee, 1962d). Moreover, many representatives highlighted the economic disparity between developed and developing nations as a significant barrier, urging specialised agencies to provide technical and financial assistance (UN, Official records of the General Assembly, Seventeenth Session, First Committee, 1297th Meeting, 1962b). Legal frameworks and principles were also a significant focus, with countries like India and Peru stressing the application of international law to outer space to ensure its peaceful use and to prevent national appropriation (UN, Official records of the General Assembly, Seventeenth Session, First Committee, 1290th Meeting, 1962a).

In 1963, during the meetings of COPUOS, the UAR proposed a draft code that highlighted international cooperation and the peaceful return of space vehicles, advocating for the involvement of developing countries in space research with support from more advanced nations (UN, Official records of the Committee on the Peaceful Uses of Outer Space, 1963a). India supported principles of non-aggression and non-appropriation, stressing that outer space should benefit all humankind (UN, Report of the Committee on the Peaceful Uses of Outer Space, 1963b). Sudan highlighted the importance of space activities in aiding all countries and welcomed initiatives for technical assistance to improve communication infrastructure (UN, Resolution 1802 (XVII), 1962c). Brazil again suggested the establishment of international sounding rocket facilities under UN auspices, emphasising technical and financial collaboration between developed and developing countries (this view would be repeated in future session also, however, it shall not be mentioned again to prevent repetition) (UN, Official records of the General Assembly, Eighteenth Session, First Committee, 1963c). Mexico and Ceylon called for comprehensive legal frameworks governing space activities, including liability for space objects and provisions for the rescue of astronauts, ensuring the protection of developing countries' interests (UN, Official records of the Committee on the Peaceful Uses of Outer Space, 1963a). Countries like Ghana and Indonesia strongly opposed the militarisation of outer space, advocating for international agreements to maintain space as a domain of peace and cooperation (UN, Report of the Committee on the Peaceful Uses of Outer Space, 1963b).

It should be noted that the focus for most nations was not on celestial property rights but on other domains in space. However, when on 13 December 1963, the Declaration of Legal Principles was established, its third principle barred national appropriation in outer space and celestial bodies (UN, Resolution 1962 (XVIII): Declaration of legal principles governing the activities of states in the exploration and use of outer space, 1963d).

In 1964, the First Committee and COPUOS deliberated on its next steps, with many members assuming that agreements on astronaut rescue or liability conventions would be prioritised. While third-world countries did make their presence known, the focus again remained on issues other than celestial property rights.

However, one important development was that the UAR referenced the Antarctic Treaty as a model for demilitarising space, indicating that a similar approach could be applied to outer space to prevent an arms race (UN, Official records of the Eighteenth Session, First Committee, 1343rd Meeting, 1963e). This is of particular note as the Antarctic Treaty bars claims of sovereignty in the icy continent whilst 'freezing' the existing claims in time (Sharma, 2024). In any case, the references to complete disarmament in space were prevalent, emphasising the need to prohibit the militarisation of outer space and to ensure that space activities were conducted for peaceful purposes.

In 1965, at the COPUOS meetings, the inputs of third world countries in the drafting of space law were significant and multifaceted. The developing countries emphasised the importance of international cooperation in space activities to ensure that the benefits of space exploration would be shared globally and used for peaceful purposes. This included advocating for frameworks that facilitate capacity-building in developing nations through education, training, and technology transfer programs, enabling them to participate actively in space activities and benefit from advancements in space science and technology. Countries such as India, Egypt, Brazil, and the UAR were active in these discussions (UN, Report of the Committee on the Peaceful Uses of Outer Space, 1966a; UN, Report of the Committee on the Peaceful Uses of Outer Space, 1965a). Third-world countries also supported the establishment of comprehensive liability frameworks to ensure that states launching space objects would be held accountable for any damage caused by those objects. This emphasis on absolute liability was seen as essential to protect less powerful states from bearing the burden of space-related accidents or mishaps (UN, Report of the Legal Sub-Committee on the work of its fourth session, 1965b). However, again, celestial property rights took a back seat.

During that year's First Committee meetings, India and the UAR stressed the importance of reserving outer space exclusively for peaceful purposes and advocated for the establishment of legal principles to prevent its militarisation, independent of broader disarmament agreements (UN, Official records of the Twentieth Session, First Committee, 1421st and 1422nd Meeting, 1965c). Ultimately, the General Assembly adopted a resolution urging the Committee to, among other things, develop international agreements on assistance to and return of astronauts and space vehicles and on liability for damage caused by objects launched into space (UN, GA Resolution on International co-operation in the peaceful uses of outer space, 1965d).

By 1966, an urgency had developed for the need for an international agreement on the exploration of the Moon and other celestial bodies (Dembling & Arons, The evolution of the Outer Space Treaty, 1967). On 7 May 1966, President Lindon B. Johnson of the US, in his statement, urged that a UN-led international treaty should become a reality (Johnson, 1966). The final agreement on the treaty resulted primarily from the efforts of the twenty-eight member Legal Subcommittee (LSC) of the UNGA COPUOS. This subcommittee held its fifth session in Geneva from 12 July to 4 August 1966 and in New York from 12 September to 16, 1966. The remaining issues were resolved through bilateral negotiations and discussions during the

Twenty-First Session of the General Assembly, leading to the announcement on 8 December that the treaty had been finalised.

Much of the existing literature on this topic primarily emphasizes the roles of the US and the USSR in the passage of the treaty, often offering little to no acknowledgment of the contributions of the Global South (Dembling & Arons, The evolution of the Outer Space Treaty, 1967; Mani et al., 1997; Gangale, 2009; Jakhu & Pelton (Jr), Global Space Governance: An International Study (Space and Society), 2017; Kemp, 1966; Lele, 2017). However, the third world played a proactive role during the final stages of the drafting of the OST (Buono, 2020). For example, on 12 July 1966, the Indian representative, Mr. Krishna Rao, speaking on behalf of the NAM nations, emphasized that these countries had consistently advocated for the LSC to work towards a convention that would ensure the peaceful use of outer space for the benefit of all humanity (UN, UNOOSA, 1966c). The NAM nations expressed satisfaction that the major space powers, the US and the USSR, had recognized the urgent need for such a convention.

However, the NAM countries also emphasised that while an agreement between the US and the USSR was crucial for progress, the interests of all states, particularly developing countries, should not be overlooked. For the Indian delegation, the immediate issue was determining the basic legal principles to apply to outer space activities. There was a debate on whether existing terrestrial laws could be applied to outer space or if entirely new concepts were needed. Ambassador Rao made reference to Mr. Andrew Haley, who, in his book "Space Law and Government," had argued that the boundaries of terrestrial jurisdiction should be clearly defined and that beyond those limits, the principle of *res communes*, or common ownership, should prevail in outer space.

On 8 December 1966, it was announced that the twenty-eight member states of the UN Outer Space Committee had reached an agreement on the text of a treaty outlining principles for the activities of states in the exploration and use of outer space, including the Moon and other celestial bodies (Dembling & Arons, The evolution of the Outer Space Treaty, 1967). The Political Committee of the General Assembly unanimously recommended approval of the treaty on 17 December 1966, and two days later, the UNGA endorsed it with a unanimous vote (Dembling & Arons, The evolution of the Outer Space Treaty, 1967).

Even after the passage of the treaty, the third world countries still made their presence felt by taking a more holistic view. Specifically, the representative from the UAR proposed the twenty-eight-power draft resolution for an international conference on the peaceful uses of outer space, emphasising the need for all countries, regardless of their development level, to benefit from space exploration. They stressed the importance of international cooperation in space activities and the necessity of space law to ensure global benefit (UN, 1491st Meeting of the First Committee, 1966b).

The OST lacks a clear and definitive framework regarding celestial property rights. While it proclaims outer space and the Moon as the "province of all mankind," promoting free and non-discriminatory access for all states, it does not explicitly adopt the concept of *res communes*. Although the treaty prohibits national sovereignty claims over outer space and celestial bodies, it leaves room for resource utilisation by any nation with the capability to access these resources. However, as highlighted in the preceding research, the Global South began to advocate for the CHM principle, taking initial steps toward a more equitable approach to space resources. This evolving perspective became more nuanced and strategically articulated during the negotiations of the 1979 Moon Agreement, as discussed in the following section.

11.5 The Moon Agreement and Third World Countries

The Moon Agreement was deliberated and developed by the LSC from 1972 to 1979 and was adopted by the General Assembly in 1979 through resolution 34/68 (UNOOSA, Agreement Governing the Activities of States on the Moon and Other Celestial Bodies, 2024). However, it was only in June 1984 that Austria, the fifth country, ratified the Agreement, leading to its enforcement in July 1984 (Christol, 1985). This Agreement reaffirms and expands upon many provisions of the OST as they pertain to the Moon and other celestial bodies, stipulating that these bodies must be used solely for peaceful purposes and that their environments must remain undisturbed. Additionally, Article 9 of the Moon Agreement requires that the UN be informed about the location and purpose of any station established on these bodies. Article 11 declares that the Moon and its natural resources are the common heritage of mankind and calls for the creation of an international regime to govern the exploitation of such resources once it becomes feasible. It is necessary to look at the provisions of this treaty as it explicitly deals with the central theme of this chapter, i.e., celestial property rights.

11.5.1 Genesis of the Moon Agreement: A Brief Synopsis

The genesis of the Moon Agreement can be dated to 1969 and to a 1970 proposal of Argentina. However, not much work was done on it until 1971. In a letter dated 27 May 1971 to the UN Secretary-General, the USSR requested the inclusion of preparing an international treaty concerning the Moon in the agenda of the twenty-sixth UNGA session, attaching the draft treaty to the letter. This item was added to the GA's agenda in 1971 and forwarded to the First Committee for consideration, which proposed a draft resolution tasking COPUOS and the LSC with drafting the treaty. The UNGA adopted this resolution on 29 November 1971 (UN, UNGA Res. Preparation of an international treaty concerning the Moon, 1971a). At its eleventh session in 1972, the LSC added "Questions relating to the moon" as a distinct agenda item. A Working Group was established to review the USSR's draft treaty and other proposals from various states, producing a draft treaty approved on 4 May 1972 as a basis for further work (UN, LSC Report – 11th Session, 1972c).

11.5.2 Drafting of the Moon Agreement and Third World Countries

In 1971, the focus of the First Committee remained on the Liability Convention (UN, Travaux préparatoires of the Agreement Governing the Activities of States on the Moon and Other Celestial Bodies, n.d.). However, the undercurrents of the Soviet draft Moon Agreement could already be felt. India welcomed the Soviet introduction of the draft treaty concerning the moon (UN, First Committee GA Verbatim Records – 26th Session, 1971b). Further, it emphasised that the resources of the Moon "must be regarded as the common heritage of mankind and not be subject to appropriation, by any party, by occupation or by extraction as property" (UN, First Committee GA Verbatim Records – 26th Session, 1971b). Ghana also supported the drafting of a new treaty once the Liability Convention had been finalised. Support was also extended by Chile, Egypt, Iran, Jamaica, Kenya, Liberia, Madagascar, Tanzania and Yugoslavia (the UAR disbanded in 1971, and Egypt took its place in the negotiations). However, Morocco, Nigeria, Indonesia and Iraq did not make any reference to the Soviet draft and instead mentioned that registration of space objects and matters relating to remote sensing was the priority (UN, First Committee GA Verbatim Records – 26th Session, 1971b).

During the meeting of the COPUOS, the Indian delegation mentioned that the benefits from lunar and celestial resources should be shared with the entire global community. However, the term "common heritage" was not utilised at this meeting: "In drawing up any treaty in respect of the Moon, it has to be ensured that the *international community is not deprived of the advantage deriving from the exploration and exploitation of the Moon*." (Emphasis supplied) (UN, UNCOPUOS Sub-Committees Verbatim Record of the Hundredth Meeting, 1971c).

The work on the Moon Treaty began in earnest in 1972. On 7 September 1972, the COPUOS reconvened for its 113th meeting. During these meetings, while the prevailing question was whether the treaty should apply to all celestial bodies or only the Moon, delegates still made reference to the celestial property rights issues. In fact, this time, the Indian delegation directly mentioned that the moon and other celestial bodies should be considered CHM (UN, COPUOS Verbatim Records, 1972a). This view was also supported by Egypt and Iran (UN, COPUOS Verbatim Records, 1972a).

Notably, it was the US that introduced the term "Common heritage of mankind" into the draft treaty (UN, COPUOS Verbatim Records, 1972a). It even envisaged "international sharing arrangements" when the exploitation of celestial resources became a possibility. However, while the Americans envisaged a sharing arrangement, it appears that the idea of the 'international regime', which would allow each country to access celestial resources, was an innovation of the third world.

On 18 October 1972, the Indian delegation, led by Mr. K.C. Pant, then Minister of State for Home Affairs, expressed satisfaction with the progress made on the Moon Treaty by the Legal Subcommittee (LSC) of the Committee on the Peaceful Uses of Outer Space (COPUOS). The delegation strongly advocated for the principle that the natural resources of the Moon should be recognized as CHM and emphasized that the proposed treaty should extend to other celestial bodies as well (UN, First Committee GA Verbatim Records – 27th Session, 1972b). The Indian delegation's stance on CHM was pivotal in promoting equitable access to and sharing of benefits from space exploration. By advocating this principle, India underscored the necessity of international agreements that would prevent monopolization of lunar resources by any single nation or technologically advanced power. India's call for CHM was not isolated; it received robust support from Afghanistan, Egypt, Ivory Coast, Jamaica, Kuwait, Sudan, Thailand, Yemen, and Yugoslavia, demonstrating a cohesive stance among developing nations towards fostering fairness and equality in the use of outer space resources (UN, First Committee GA Verbatim Records – 27th Session, 1972b).

On 25 June 1973, a new session of the UN COPUOS began. On 28 June, India reiterated that it considered the Moon, other celestial bodies and their resources as CHM. This time, the statement went a bit further: "We wish also to reiterate that the treaty concerning the moon should undertake an obligation to inform the international community of any indication of natural resources on the moon that they might discover in the course of their exploitation." (UN, COPUOS Verbatim Records, 1973).

Adding to its previous stance, this statement emphasised the need for transparency and international cooperation in lunar exploration. It called for a treaty obligation to inform the global community about any natural resources discovered on the moon during exploitation. This would ensure that all nations are aware of such findings, promoting shared knowledge and equitable access to lunar resources. Ultimately, the meeting resulted in a draft agreement with 21 articles and a preamble.

In 1974, the focus of COPUOS remained on registration convention and remote sensing. Thus, not much meaningful discussion was done on the Moon Agreement. However, in the work of the LSC that year, the 'Draft treaty relating to the moon' was adopted as the second agenda. India, Egypt, and Nigeria co-sponsored (both together and separately) working papers on the articles of the draft treaty (UN, LSC Report – 13th Session, 1974a). However, the LSC meetings provided a different picture. The meetings were held between 6 to 31 May, 1974 (UN, Legal Sub-Committee – Summary Records of the Two Hundred and Eighth to Two Hundred and Twenty-Fifth Sessions, 1974b). The draft Moon Agreement remained an issue to yet be reckoned with. The Chairman himself noted that the most difficult question with respect to the draft Moon Treaty remained the question of natural resources. For India, the Moon, celestial bodies and their resources were CHM and this should had to be formally declared. Among other things, the agreement should ensure equitable sharing of resources, taking into concern the needs of the developing countries. This view was supported by participating third world countries, including Egypt, Iran, and Nigeria. However, the discussion was to continue as no final draft was reached.

The fourteenth session of the LSC lasted from 10 February 1975 till 7 March 1975 (UN, Legal Sub-Committee – Summary Records of the Two Hundred and Twenty-Sixth to Two Hundred and Forty-Fifth Sessions). In the sessions, India reiterated its position from the previous years. Argentina, Iran, and Venezuela supported this view (UN, Legal Sub-Committee – Summary Records of the Two Hundred and Twenty-Sixth to Two Hundred and Forty-Fifth Sessions, 1975a). Egypt supported the CHM view in principle. Notably, in a rare show of comradery, Pakistan explicitly supported India's submission: "..most of the developing countries were poor and also anxious to improve the quality of life in their countries. They could not afford to invest their resources in a space race..(Pakistan) felt that the errors which had led to colonialism must not be repeated in outer space and that the principle of the common heritage of mankind would avert a repetition of the follies of the past..(Pakistan's) delegation endorsed the appeal made by the Indian representative not to hem the common heritage principle.." (UN, Legal Sub-Committee – Summary Records of the Two Hundred and Twenty-Sixth to Two Hundred and Forty-Fifth Sessions, 1975a).

From the above, it would appear that Pakistan adopted a 'Third World' approach, which upended its generally established diplomatic position vis-à-vis India. However, the matter still remained in limbo. Later on, during the general debate of the COPUOS in June 1975, Iran and Kenya were the only third-world countries that highlighted the Moon Agreement. In Kenya's opinion at the time, the concept of "common heritage of mankind applies mutatis mutandis to the moon and other celestial bodies" (UNGA, COPUOS Verbatim Records, 1975). During the session on 11 June 1975, Iran's delegation succinctly summarised the situation as follows: "In reviewing the work of the Legal Sub-Committee, we once more find ourselves disappointed at the lack of progress in connection with the Moon Treaty. Despite the time and effort expended on this subject, we find that the deadlock which developed in 1973—after the text of 21 articles and the preamble had been agreed upon—still persists. The main issue which has resisted solution still revolves around the question of the legal regime of the natural resources of the Moon." (UN, COPUOS Verbatim Records, 1975b).

The LSC reconvened on 3 May 1976. Similar to the previous years, third world countries such as Chile, Egypt, India, Mexico and Sierra Leone were of the opinion that the natural resources on the Moon and other celestial bodies shall be CHM (UN, LSC Summary Records – 15th Session, 1976a). In the meeting on 23 June 1976, Chile, recalling the Convention on the Law of the Sea, which was being drafted at the time, made a pertinent observation: "Our delegation is convinced that there are no valid reasons for not applying to the natural resources of celestial bodies, and in particular of the Moon, the same regime of 'common heritage of mankind' which has been established for the sea bed beyond the limits of national jurisdiction. Such resources should not be the property only of those who are able to explore and exploit them; rather, they should be distributed for the benefit of all mankind without discrimination whatsoever, and States should commit themselves to establishing an international regime and procedure to regulate those activities" (UN, COPUOS Verbatim Records, 1976b). This view is a different understanding of CHM and reflective of the thinking of the developing countries and foreshadows the differences of interpretation that would arise later within the Moon Agreement.

The year 1977 was significant as it heralded the tenth year anniversary of the OST. From 14 March 1977, the work on the Moon Treaty gained prominence. A working group had already been established by the LSC to deal with the treaty. Further, the working group had also decided to "give priority to the question of natural resources, since many delegations felt that if that issue were resolved, it would be easier to reach an agreement on the remaining two issues" (UN, LSC Summary Records – 16th Session, 1977a). During the general exchange of views, Chile made a statement akin to its view in the previous year. India, Indonesia, and Iran stated that they considered lunar resources CHM (UN, LSC Summary Records – 16th Session, 1977a). As per procedure, the report of the LSC was then considered by the COPUOS in its 20th session between 20 June 1977 and 1 July 1977. Chile, Egypt, India, Indonesia, Mexico and Nigeria were again of the opinion that the CHM principle cannot be alienated from the treaty (UN, COPUOS Verbatim Records, 1976c).

Even though, again, no new treaty had been drafted for the Moon, there was a refinement of the idea of CHM and its greater acceptability within third-world countries. By 1977, it became clear that the third world's approach towards celestial and lunar resources would be based on *res communis.* At the same time, there were also calls for the establishment of an international body to regulate and distribute the profits from the use/acquisition of the said resources. Ultimately, the UNGA recommended that the LSC continue its consideration of the draft treaty relating to the moon on a priority basis (UN, UNGA Res on International co-operation in the peaceful uses of outer space, 1977b).

In 1978, the meetings of the LSC lasted from 13 March 1978 till 7 April 1978. The first week of the LSC deliberations was dedicated to the draft treaty relating to the moon. In furtherance of this, the Working Group I on the draft treaty held seven meetings (UN, LSC Report – 17th Session, 1978a). In the beginning, the two major space powers did not concede their intentions. While the USSR stated that the most acceptable draft to all nations should be selected, the US chose to maintain ambiguity by stating that: "The United States delegation was prepared to co-operate with other delegations to reach a consensus (regarding natural resources of the moon and other celestial bodies) which would permit the Sub-Committee to complete the drafting of the treaty" (UN, LSC Summary Records – 17th Session, 1978b). Such views were shared by other countries such as Austria and Poland (UN, LSC Summary Records – 17th Session, 1978b).

On the other hand, Indonesia again reiterated the views of the third world: "Any natural resources which might be found on the moon could therefore be commercially exploited only in the context of an agreed international regime, which would guarantee not mainly the opportunity for the exploitation and use of such resources, but also the equitable sharing of the benefits obtained. The principle of the common heritage also precluded any claim of sovereignty by a nation or group of nations either over the moon itself or over any other celestial body. Hence, logically, there could be no claim to their natural resources."

Its delegate further said that: "The principle of the common heritage of mankind should be explicitly recognised, as the developing countries had proposed. Although it would be a long time before the exploration and exploitation of the moons, resources became a reality, it was important to map out the political and legal framework now, in audit, to ensure that space technology benefited all once practical applications were possible. Any possible use of the resources exceeding the limits of mere exploration should be undertaken on an agreed international basis, and the benefits should be shared equally among all." (UN, LSC Report – 17th Session, 1978a).

Colombia's stance is also pertinent to the discussion: "the initial inequality between States should be taken into account that they should not all be treated in the same way. The formula 'common heritage of mankind' implied that the methods hitherto commonly employed on the Earth should be replaced by a spirit of solidarity in the exploitation of outer space." (UN, LSC Summary Records – 17th Session, 1978b). The views of other nations are tabulated below in Table 11.1:

It should be highlighted here that most delegations in favour of the CHM principle were the third-world nations. Of course, there were outliers, such as the Netherlands and Turkey. Other countries did not either mention the issue or suggested an alternative. However, a working paper submitted by Austria became a pertinent basis for a final draft in 1979 (UN, Annex I, Appendix, 13 April, 1978). Due to the paucity of time in 1978, the draft could only be discussed in informal consultations (UN, LSC Report – 18th Session, 1978c). The draft is quite similar to the final treaty, which will be finalised in the 18th session of the LSC.

Once the LSC meetings concluded, its report was forwarded to the COPUOS as per procedure. The COPUOS held its twenty-first session at the UN Headquarters from 26 June to 7 July 1978 (UN, Report of the Committee on the Peaceful Uses of Outer Space – 21st Session). Due to events at the time, the focus of most countries was on other issues, such as direct television broadcasting and the use of artificial Earth satellites. Still, many commented on the Moon Agreement and the celestial bodies (see Table 11.2 below):

Just like in previous years, the UNGA passed a resolution that the LSC continue work on the draft treaty relating to the Moon (UN, GA Res 33/16, GAOR, 1978). However, in 1979, an agreement would be reached.

In 1979, work on the Moon Agreement started for the last time in the 18th LSC session held from 12 March to 6 April 1979. The sub-committee noted that it had in 1972 approved the texts of a preamble and 21 articles, including final clauses, and that in 1973, it had taken note of the text of six provisions which were formulated that year in its working group. It had not as yet succeeded, however, in resolving the three main outstanding issues:

1. The scope of the treaty
2. The information to be furnished on missions to the moon and
3. The natural resources of the moon.

The question concerning the natural resources of the moon was generally recognised to be the key issue whose solution could facilitate agreement on the two other remaining issues (UN, LSC Report – 18th Session, 1979a). Further, the LSC also noted that the draft agreement submitted by Austria in the previous session could serve as a basis for finalising the preparation of an international instrument relating to the moon and other celestial bodies. The LSC report shows that a compromise

Table 11.1 List of countries and their views regarding celestial resources (17th Session, 1978)

S.No.	Country	View Regarding Celestial Resources	Document Code
1.	Soviet Union	Not addressed	A/AC.105/C.2/ SR.285
2.	Canada	Not addressed	-do-
	Japan	Not addressed	-do-
3.	Austria	Not addressed	A/AC.105/C.2/ SR.288
4..	Poland	Not addressed	-do-
5.	United States	TBD after discussion with all States Parties	-do-
6.	Indonesia	CHM	-do-
7.	United Kingdom	Not addressed	-do-
8.	Philippines	CHM	-do-
9.	Turkey	CHM	-do-
10.	Yugoslavia	Not addressed	-do-
11.	Chile	CHM	A/AC.105/C.2/ SR.289
12.	Mexico	CHM	-do-
13.	Italy	Equitable sharing of resources between Space Powers and Developing Nations	-do-
14.	Sierra Leone	CHM	-do-
15.	German Democratic Republic	Celestial resources can be mentioned in a protocol rather than the treaty	-do
16.	Australia	CHM is doubtful due to different interpretations; a more direct approach is required	-do-
17.	Nigeria	Not addressed	A/AC.105/C.2/ SR.290
18.	India	CHM	-do-
19.	Czechoslovakia	Not addressed	-do-
20.	Brazil	CHM	-do-
21.	Netherlands	CHM	-do-
22.	Canada	Not addressed	-do-
23.	Mongolia	Celestial resources (including CHM) can be mentioned in a protocol rather than the treaty	-do-
24.	Egypt	CHM	-do-
25.	Iran	CHM	-do-
26.	France	Not addressed	-do-
27.	Kenya	CHM	A/AC.105/C.2/ SR.291
28.	Argentina	CHM	-do-
29.	Bulgaria	CHM is doubtful due to different interpretations; different approach is required	-do-
30.	Colombia	CHM	-do-
31.	Japan	Not addressed	-do-

(continued)

Table 11.1 (continued)

S.No.	Country	View Regarding Celestial Resources	Document Code
32.	Hungary	Not addressed	-do-
33.	USSR	Not addressed	-do-
34.	US	Not addressed	-do-

Table 11.2 List of countries and their views regarding celestial resources (COPUOS 21st Session)

S.No.	Country	View Regarding Celestial Resources	Document Code
1.	Poland	Not addressed	A/AC.105/PV.181
2.	Canada	Compromise required for resource exploitation to finalise treaty	-do-
3.	Netherlands	CHM	-do-
4.	German Democratic Republic	Not addressed	-do-
5.	Sweden	Not addressed	-do-
6.	Argentina	Not addressed	-do-
7.	Ecuador	CHM	-do-
8.	USSR	Supported CHM principle as a basis for compromise	-do–/ A/AC.105/ PV.185
9.	Czechoslovakia	Not addressed	A/AC.105/PV.181
10.	Chile	CHM	A/AC.105/PV.184
11.	Iran	CHM	-do-
12.	Colombia	Not addressed	-do-
13.	Indonesia	CHM	-do-
14.	Mongolia	CHM	-do-
15.	Philippines	Not addressed	-do-
16.	Yugoslavia	Not addressed	-do-
17.	Kenya	Not addressed	-do-
18.	Australia	Not addressed	A/AC.105/PV.185
19.	Romania	Not addressed	-do-
20.	India	Not addressed	-do-
21.	Sudan	CHM	-do-

was reached through the working paper submitted by Austria in 1978. As stated above, the draft could not be discussed properly in 1978 due to lack of time. However, at that time, the LSC was hopeful that the draft would facilitate reaching a consensus.

In 1979, the LSC began its discussions with the Austrian draft on the centre stage. The working group, by consensus, decided to consider each draft article one by one. With reference to natural resources, some delegations enunciated that the moon and other celestial bodies (and their natural resources) should be considered as the CHM. Further, it was also posited that an international regime governing the exploitation of the moon's natural resources should be established. On the other hand, other delegations wanted to limit the scope of the treaty to the moon only, and due to the contemporary inability to access the resources on the moon, they did not want any provisions for the natural resources of the moon.

After the discussion, the "Working Paper Reflecting the Outcome of the Review at the Present Session of Working Paper WG. I (1978)/WP.2 of 3 April 1978" was generated (UN, LSC Report – 18th Session, 1979a). It was titled the "Agreement Governing the Activities of States on the Moon and other Celestial Bodies" (UN, LSC Report – 18th Session, 1979a).

Broadly speaking, it appears that in 1979, during the LSC meetings, the developing world maintained strong support for the CHM principle in the context of the Moon Agreement. Following these discussions, the COPUOS convened from 18 June to 3 July 1979 (UN, COPUOS Report – 34th Session, 1979c). At this stage, the draft agreement was ready for the Committee's consideration. However, for reasons that remain unclear, the draft treaty was unexpectedly shelved. This sudden halt in progress took many third-world countries by surprise. On 18 June 1979, Pakistan, while acknowledging the effots of the LSC, noted with concern that the "draft treaty relating to the moon and other celestial bodies was proposed by the Austrian delegation at the seventeenth session of the Legal Sub-Committee, but no significant progress could be made during the consideration of this draft at the twenty-first session of the Committee" (UN, COPUOS Verbatim Records, 1979b). Similarly, Egypt and India expressed disappointment over the lack of widespread acceptance of the Austrian draft (UN, COPUOS Verbatim Records, 1979b). The Indian delegation remarked, "Finally, with regard to the draft treaty relating to the moon, very commendable work has been done by the Austrian delegation. We had hoped that during the last session of the Legal Sub-Committee, that draft would have been generally accepted and approved. We had decided, after a lot of thinking, that we would go along with it. *But something seems to have happened. I do not know what*" (UN, COPUOS Verbatim Records, 1979b) [Emphasis supplied].

The delegation then went on to express hope that the draft agreement would be resurrected later; otherwise, the clock on the work would turn back by a year. As the deliberations went on, the Indonesian delegation also supported the Austrian proposal and hoped that it would become a basis for reaching a consensus in the present session (UNGA, COPUOS Verbatim Records, 1979). Ecuador and Romania again emphasised that CHM has to be a part of any agreement regarding the Moon, along with an international regime to manage its resources. Turkey also said that the countries should have a Right to benefit from the resources of the moon on a fair and equitable basis (UNGA, COPUOS Verbatim Records, 1979). Chile mentioned that it had been optimistic that the Austrian draft would work in the LSC; however, the same did not occur. The delegation was now hopeful that an acceptable treaty would be able to come out of the COPUOS informal working group (UNGA, COPUOS Verbatim Records, 1979).

The US also noted that the Austrian text was a "meritorious one, representing a meritorious effort" (UNGA, COPUOS Verbatim Records, 1979). The United Kingdom also viewed the Austrian draft as a worthy basis for an acceptable Moon treaty. Australia was also taken aback by the non-acceptance of the Austrian text and wondered whether it would be possible to even conclude a treaty this time. On the other hand, the German Democratic Republic mentioned that it might be advisable in the view of divergent views of members to no longer give high priority to the issue. The Japanese also expressed their disappointment with the work of the LSC and advised that further efforts should be made in the next session.

Representatives of Argentina, Australia, Austria, Belgium, Brazil, Bulgaria, Canada, Chile, Colombia, Czechoslovakia, Ecuador, Egypt, France, German Democratic Republic, Germany, Hungary, India, Indonesia, Iran, Iraq, Italy, Japan, Kenya, Mexico, Mongolia, Netherlands, Nigeria, Pakistan, Philippines, Poland, Romania, Sierra Leone, Sweden, Turkey, USSR, US, Venezuela and Yugoslavia attended the meetings. The Committee considered the compromise text that had been proposed by Austria. Subsequently, Article XI, paragraph 1 of the Austrian text was amended as follows: "The moon and its natural resources are the common heritage of mankind, which finds its expression in the provisions of this agreement and, in particular, in paragraph 5 of this article."

This amendment was one of several changes made to the agreement and holds particular relevance to the present research. Ultimately, COPUOS reached a consensus on the final text of the Moon Agreement. Notably, the success of the third-world countries in advocating for terms that did not solely favour Western interests marked a significant achievement. Their persistent efforts ensured that the treaty incorporated principles of equity and fairness, reflecting a more balanced approach to international space law.

11.6 Conclusion

The above research shows that third-world nations have had a nuanced approach to ISL from its very inception. It would appear that at the beginning, when the OST had to be drafted, the nations focused on peace and, at the same time, other space applications. Even before that, when the composition of the *ad hoc* Committee was being looked into, some third-world countries, such as India and the UAR, instead focused on the creation of a holistic body that had the presence of both superpowers of the time. This seems to have been a non-aligned approach where one bloc was not being supported against another. Obviously, the non-aligned approach became especially evident after NAM was formulated.

Once COPUOS had been created, the third-world countries made sure that their views were put on the record. This is evident in the research done above. However, at this time, the focus was not categorically on celestial property rights but rather on making sure that the Cold War did not slip into outer space. Due to this, they seem to have agreed to the notion that outer space is a province of mankind rather than a common heritage. However, the views of the third world become apparent during the negotiations of the Moon Agreement of 1979. These countries wanted the creation of a regime that would declare the resources on the moon as CHM. They also wanted the idea of global commons to apply to space and such celestial bodies. Further, they also demanded the extension of the application of this agreement to other celestial bodies, i.e., not limited only to the moon. At the same time, they wanted to focus on the creation of an international regime that would equitably share the resources attained from the moon with all the nations.

The third-world approach can be construed from the deliberations that took place during the COPUOS meetings. For instance, during the LSC deliberations in 1977, the UK said it was flexible in finding an approach towards resource governance on the Moon, whereas the French were of the opinion that it was premature to discuss the topic itself (UN, LSC Summary Records – 16th Session, 1979a). This can be differentiated from the approach of third world nations, which, even with minor differences, was quite streamlined.

The concept of the CHM gained relevance through the principles of international cooperation and shared benefit enshrined in the Antarctic Treaty of 1959. While the treaty did not explicitly adopt the CHM principle, it set a precedent for collective stewardship by designating Antarctica as a region reserved for peaceful purposes and scientific research for the benefit of all humanity. The CHM principle was more formally articulated during the negotiations of the United Nations Convention on the Law of the Sea (UNCLOS) in 1982, where the deep seabed was explicitly designated as the "common heritage of mankind." Developing countries played a pivotal role in advocating for this regime, emphasizing equitable resource sharing and preventing monopolization by technologically advanced states. Even the UNCLOS came after the Moon Agreement, the CHM principle had already been articulated in the context of the peaceful uses of the sea-bed and the ocean floor by Arvid Pardo, Malta's ambassador to the United Nations in the year 1967. He proposed that the deep seabed should not be subject to national appropriation, laying the groundwork for CHM in both maritime and space law contexts (UN, General Assembly, 22nd session: first Committee, 1516th meeting, Wednesday, 1 November 1967, New York, 1967) Building on this momentum, these nations sought to extend the CHM principle to outer space during the drafting of the Moon Agreement in 1979. Their objective was to ensure that celestial bodies, including the Moon, would be governed by the same equitable and inclusive framework, promoting global benefit and preventing new forms of imperialism in the emerging domain of space exploration.

The Austrian proposal emerged as a critical starting point in the negotiations, undergoing multiple amendments throughout the drafting process. One of the most significant challenges during these negotiations was determining the scope of the treaty, that is, whether it would apply solely to the Moon or extend to other celestial bodies as well, and establishing clear guidelines for the utilization of lunar resources. Between 1972 and 1978, developing countries consistently championed the CHM principle, advocating for a legal framework that would prevent the monopolization of extraterrestrial resources. Their persistence paid off during the final phase of negotiations when the CHM principle was formally incorporated into the treaty, gaining approval from the UN General Assembly. For the Global South, the notion of celestial property ownership was never an acceptable outcome. Instead, they focused on ensuring that any resources extracted from the Moon would be accessed and shared equitably. The underlying principle was clear: regardless of a nation's technological capabilities, any benefits derived from lunar resources should be distributed fairly among all countries. This commitment to equity and inclusivity is a cornerstone of the Moon Agreement, reflecting the strategic and principled stance of third-world countries in promoting a more just and balanced approach to space governance.

In international law, no country can be compelled to adhere to a particular legal regime. The principle of state sovereignty allows nations to freely decide which treaties they wish to ratify, regardless of their participation in the treaty's formation and drafting processes. This autonomy is a double-edged sword, as demonstrated by the Moon Agreement's limited success compared to its 1967 sibling, the OST. While the latter enjoys broad acceptance, the former has only 17 parties as of 2025, reflecting its struggle to gain widespread international support. This discrepancy underscores the challenges of establishing binding international norms, particularly when powerful spacefaring nations choose to remain outside the treaty framework. Nevertheless, this research highlights that Third-World countries maintained a distinct perspective on celestial property rights, grounded in the CHM principle. Their advocacy for CHM was not merely aspirational but represented a strategic effort to embed equity and shared benefits into International Space Law. This stance aligns with a Third World Approach to International Law, emphasising the need to rectify historical inequalities and promote a more inclusive and just global legal order, particularly in the new frontier of outer space exploration.

References

Aganaba-Jeanty, T. (2016). Introducing the Cosmopolitan Approaches to International Law (CAIL) lens to analyze governance issues as they affect emerging and aspirant space actors. *Space Policy*, 3–11.

Al Attar, M. (2020). TWAIL: A Paradox within a Paradox. *International Community Law Review*, 163–196.

al Attar, M. (2021). Must international legal pedagogy remain eurocentric? *Asian Journal of International Law*, 176–206.

Anghie, A. (2007). *Imperialism, sovereignty and the making of international law*. Cambridge University Press.

Anghie, A. (2023). Rethinking international law: A TWAIL retrospective. *European Journal of International Law*, 7–112.

Appiagyei-Atua, K. (2015). Ethical Dimensions of Third-World Approaches to International Law (TWAIL): A critical review. *African Journal of Legal Studies*, 209–235.

Bhat, S. (2009). *Outer space law: From theory to practice*. ICFAI University Press.

Buono, S. (2020). Merely a 'scrap of paper'? The Outer Space Treaty in historical perspective. *Diplomacy & Statecraft, 31*, 350–372.

Burra, S. (2016). TWAIL's others: A caste critique of TWAILers and their field of analysis. *Windsor YB Access Just, 111*.

Cheng, B. (1997). *Studies in international space law*. Oxford University Press.

Chimni, B. S. (2003). Third world approaches to international law: a manifesto. In A. Anghie, B. S. Chimni, K. Mickelson, & O. C. Okafor (Eds.), *The third world and international order: Law, politics and globalization* (pp. 27–73). Brill.

Chimni, B. S. (2022). The international law of jurisdiction: A TWAIL perspective. *Leiden Journal of International Law*, 29–54.

Christol, C. Q. (1985). The Moon Treaty enters into force. *American Journal of International Law, 79*, 164–172.

COPUOS. (1962). *Report of the legal sub-committee on the work of its first session.* UNGA, COPUOS. UN.

COPUOS. (2024, April 15). *Status of International Agreements relating to activities in outer space as at 1 January 2024.* Retrieved from UNOOSA: https://www.unoosa.org/res/oosadoc/data/documents/2024/aac_105c_22024crp/aac_105c_22024crp_3_0_html/AC105_C2_2024_CRP03E.pdf

Dembling, P. G., & Arons, D. M. (1967). The evolution of the Outer Space Treaty. *Journal of Air Law and Commerce, 33*, 421–450.

Dinkel, J. (2018). *The Non-Aligned Movement: genesis, organization and politics (1927–1992).* Brill.

Emtseva, J. (2022, 02.07). *Teaching Karl Marx about third world approaches to international law.* Retrieved from Opinio Juris: https://opiniojuris.org/2022/02/07/teaching-karl-marx-about-third-world-approaches-to-international-law/

Enyew, E. L. (2022). Sailing with TWAIL: A historical inquiry into third world perspectives on the law of the sea. *Chinese Journal of International Law*, 439–497.

Falk, R. (2016). Foreword: Third World Approaches to International Law (TWAIL) special issue. *Third World Quarterly*, 1943–1945.

Fidler, D. P. (2003). Revolt against or from within the West? TWAIL, the developing world, and the future direction of international law. *Chinese Journal of International Law*, 29–76.

Gangale, T. (2009). *The development of outer space: Sovereignty and property rights in international space law*. Praeger.

Gathii, J. T. (2009). *War, commerce, and international law*. Oxford University Press.

Gathii, J. T. (2020). The promise of international law: A third world view. In *Proceedings of the ASIL annual meeting* (pp. 165–187). Cambridge University Press.

Haskell, J. (2014). Trail-ing TWAIL: Arguments and blind spots in Third World approaches to international law. *Canadian Journal of Law & Jurisprudence*, 383–414.

Jain, S., Pannu, A. K., & Godara, K. (2022). *International law*. Allahabad Law Agency.

Jakhu, R. S., & Pelton, J. N. (2017). *Global space governance: An international study*. Springer.

Jakhu, R. S., & Pelton, J. N., Jr. (Eds.). (2017). *Global space governance: An international study (space and society)*. Springer International Publishing AG.

Jessup, P. C., & Taubenfeld, H. J. (1959). The United Nations ad hoc committee on the peaceful uses of outer space. *American Journal of International Law*, 877–881.

Johnson, L. B. (1966, May 07). *Statement by the President on the need for a treaty governing exploration of celestial bodies*. Retrieved from The American Presidency Project: https://www.presidency.ucsb.edu/documents/statement-the-president-the-need-for-treaty-governing-exploration-celestial-bodies

Kapur, R. (2023). TWAIL and alternative visions: 'Talking About a Revolution': Afterword to the Foreword by Antony Anghie. *European Journal of International Law*, 771–778.

Kemp, J. M. (1966). *Evolution toward a space treaty: An historical analysis*. Washington DC.

Kennedy, J. F. (1962, September 12). *"We choose to go to the Moon"*. Retrieved October 2024, from Rice University: https://www.rice.edu/jfk-speech

Lele, A. (2017). *50 years of the Outer Space Treaty: Tracing the journey*. Pentagon Press.

Lyall, F., & Larsen, P. (2009). *Space law: A treatise*. Ashgate Publishing.

Mani, V. S., Bhatt, S., & Reddy, B. (1997). *Recent trends in international space law and policy*. Lancers Books.

Modirzadeh, N. (2023). "Let Us All Agree to Die a Littke": TWAIL's unfulfilled promise. *Harvard International Law Journal*, 79–131.

Morland, A. (2024). The outer space regime of property rights through the lens of the third world: Towards a copernican revolution from anthropocentrism to cosmocentrism. *Revue interdisciplinaire d'études juridiques*, 159–183.

Mutua, M. (2000). What is TWAIL?. *Proceedings of the ASIL annual meeting*. Cambridge University Press.

Natarajan, U., Reynolds, J., & Sujith, A. B. (2018). Introduction: TWAIL – on praxis and the intellectual Chapter Introduction: TWAIL – on praxis and the intellectual. In U. Natarajan, J. Reynolds, & A. B. Sujith (Eds.), *Third world approaches to international law: On praxis and the intellectual* (pp. 4–14). Routledge.

Obiora Chinedu, O. (2008). Critical Third World approaches to international law (TWAIL): Theory, methodology, or both? *International Community Law Review*, 371–378.

Okafor, O. (2008). Critical Third World approaches to international law (TWAIL): Theory, methodology, or both? *International Community Law Review*, 371–378.

Pahuja, S. (2011). *Decolonising international law: Development, economic growth and the politics of universality*. Cambridge University Press.

Pic, P., Evoy, P., & Morin, J.-F. (2023). Outer space as a global commons. *International Journal of the Commons*, 288–301.

Sharma, A. V. (2024). *Space tourism: Learning from the Antarctic experience*. (S. B. B, Ed.) New York: Routledge.

Steer, C. (2019). Who has the power? A critical perspective on space governance and new entrants to the space sector. *Georgia Journal of International and Comparative Law*, 751–760.

Storr, C. (2021). Space is the only way to go': The evolution of the extractivist imaginary of international law. In S. Chalmers & S. Pahuja (Eds.), *Routledge handbook of international law and the humanities*. Routledge.

Su, J. (2022). Legal status of abiotic resources in outer space: Appropriability, ownership, and access. *Leiden Journal of International Law*, 825–852.

Tronchetti, F. (2009). *The exploitation of natural resources of the Moon and other celestial bodies: A proposal for a legal regime*. Martinus Nijhoff Publishers.

UN. (1962a). *Official records of the general assembly, seventeenth session, first committee, 1290th meeting*. UNGA. United Nations.

UN. (1962b). *Official records of the general assembly, seventeenth session, first committee, 1297th meeting*. UNGA. UN.

UN. (1962c). *Resolution 1802 (XVII)*. UNGA. UN.

UN. (1962d). *Official records of the General Assembly, Seventeenth Session, First Committee, 1295th Meeting United Nations, 1962*. Official records of the General Assembly, Seventeenth Session, *First Committee*. UNGA. UN.

UN. (1963a). *Official records of the committee on the peaceful uses of outer space*. UNGA. UN.

UN. (1963b). *Report of the committee on the peaceful uses of outer space*. UNGA. UN.

UN. (1963c). *Official records of the General Assembly, Eighteenth Session, First Committee*. UNGA. UN.

UN. (1963d). *Resolution 1962 (XVIII): Declaration of legal principles governing the activities of states in the exploration and use of outer space*. UNGA. UN.

UN. (1963e). *Official records of the Eighteenth Session, First Committee, 1343rd Meeting*. UNGA. UN.

UN. (1965a). *Report of the committee on the peaceful uses of outer space*. UNGA. UN.

UN. (1965b). *Report of the Legal Sub-Committee on the work of its fourth session*. UNGA. UN.

UN. (1965c). *Official records of the Twentieth Session, First Committee, 1421st and 1422nd meeting*. UNGA. UN.

UN. (1965d). *GA Resolution on International co-operation in the peaceful uses of outer space*. UNGA. UN.

UN. (1966a). *Report of the committee on the peaceful uses of outer space*. UNGA. UN.

UN. (1966b). *1491st Meeting of the first committee*. UNGA.

UN. (1966c). *UNOOSA*. Retrieved from Report of the United Nations Conference on the Exploration and Peaceful Uses of Outer Space: https://www.unoosa.org/pdf/gadocs/A_7285E.pdf

UN. (1967). *General Assembly, 22nd session: 1st Committee, 1516th meeting*, Wednesday, 1 November 1967, New York. UNGA, First Committee. UN.

UN. (1971a). *UNGA Res. Preparation of an international treaty concerning the Moon*. UNGA. UN.
UN. (1971b). *First committee GA verbatim records – 26th Session*. UNGA.
UN. (1971c). *UNCOPUOS sub-committees verbatim record of the hundredth meeting*. UNGA.
UN. (1972a). *COPUOS verbatim records*. UNGA. UN.
UN. (1972b). *First committee GA verbatim records – 27th session*. UNGA. UN.
UN. (1972c). *LSC Report – 11th session*. UNGA. UN.
UN. (1973). *COPUOS verbatim records*. UNGA. UN.
UN. (1974a). *LSC Report – 13th session*. UNGA. UN.
UN. (1974b). *Legal sub-committee – summary records of the two hundred and eighth to two hundred and twenty-fifth sessions*. COPUOS.
UN. (1975a). *Legal sub-committee – summary records of the two hundred and twenty-sixth to two hundred and forty-fifth sessions*. COPUOS. UN.
UN. (1975b). *COPUOS verbatim records*. COPUOS. UN.
UN. (1976a). *LSC Summary Records – 15th session*. COPUOS. UN.
UN. (1976b). *COPUOS verbatim records*. COPUOS. UN.
UN. (1976c). *COPUOS verbatim records*. UNGA.
UN. (1977a). *LSC summary records – 16th Session*. COPUOS. UN.
UN. (1977b). *UNGA Res on International co-operation in the peaceful uses of outer space*. UNGA.
UN. (1978a). *LSC Report – 17th session*. UNGA. UN.
UN. (1978b). *LSC summary records – 17th session*. COPUOS. UN.
UN. (1978c). *LSC report – 18th session*. COPUOS. UN.
UN. (1979a). *LSC report – 18th session*. COPUOS. UN.
UN. (1979b). *COPUOS verbatim records*. COPUOS. UN.
UN. (1979c). *COPUOS Report – 34th session*. COPUOS. UN.
UN. (2024, November 6). *COPUOS history*. Retrieved from United Nations Office of Outer Space Affairs: https://www.unoosa.org/oosa/en/ourwork/copuos/history.html
UN. (n.d.). *Travaux préparatoires of the Agreement Governing the Activities of States on the Moon and Other Celestial Bodies*. Retrieved October 2024, from UNOOSA: https://www.unoosa.org/oosa/en/ourwork/spacelaw/treaties/travaux-preparatoires/moon-agreement.html
UNGA. (1958a, December 13). *Res. 1384 (XIII) on the Question of the peaceful use of outer space*. Retrieved October 2024, from Resolutions adopted on the reports of the First Committee: https://www.unoosa.org/pdf/gares/ARES_13_1348E.pdf
UNGA. (1958b). *Summary record of the 995th meeting of the 1st Committee*. United Nations General Assembly, First Committee. New York: UN.
UNGA. (1959a). *Report of the ad hoc committee on the peaceful uses of outer space*. United Nations, United Nations Office of Outer Space Affairs. New York: United Nations.
UNGA. (1959b). *Comments on the report of the ad hoc committee on the peaceful uses of outer space*. UN, UNGA. New York: UN.
UNGA. (1959c). *GA Res on International co-operation in the peaceful uses of outer space*. UN, UNGA. New York: UN.
UNGA. (1962a). *Report of the committee on the peaceful uses of outer space*. UNGA, Committee on the Peaceful Uses of Outer Space. United Nations.
UNGA. (1962b). *Official records of the general assembly, seventeenth session, first committee, 1297th meeting*. United Nations, First Committee.
UNGA. (1975). *COPUOS verbatim records*. COPUOS. UN.
UNGA. (1979). *COPUOS verbatim records*. COPUOS. UN.
UNOOSA. (2017). *International space law: United Nations Instruments*. Retrieved from UNOOSA: https://www.unoosa.org/res/oosadoc/data/documents/2017/stspace/stspace61rev_2_0_html/V1605998-ENGLISH.pdf
UNOOSA. (2022). *United Nations treaties and principles on outer space*. Retrieved from https://www.unoosa.org/pdf/publications/STSPACE11E.pdf

UNOOSA. (2024). *Agreement governing the activities of states on the moon and other celestial bodies*. Retrieved from United Nations Office of Outer Space Affairs: https://www.unoosa.org/oosa/en/ourwork/spacelaw/treaties/intromoon-agreement.html

van Eijk, C., & Aganaba, T. (2021). Inspired by Africa: A new approach to global space governance. *New Space*, 1–4.

Vazhapully, K. M. (2024). Third world approaches to international space law. In I. S. Bhat, D. Ukey, & A. Variath (Eds.), *International space law in the new space era: Principles and challenges* (pp. 295–320). Oxford University Press.

Index

A
African Space Policy, 142
Anti-satellite weapons (ASAT), 104, 105
Apollo, 2, 19, 127, 138, 143
Armstrong, N., 127
Astronauts, 24, 74, 100, 102, 127, 177, 182–184

C
Cold War, 19, 56, 89, 106, 116, 124, 128, 144, 169, 180, 181, 195
Commercialisation, 25, 27, 35, 54, 58, 70, 71, 80, 124, 129, 130, 140, 143, 158, 161, 165, 168, 170
Commercial Space Launch Competitiveness Act, 2015, 58, 64, 91
Commercial utilisation, 3, 166
Common heritage of mankind (CHM), 9, 10, 28, 43–45, 47, 57, 73, 74, 81, 90, 91, 131, 145, 158, 160, 162, 178, 180, 186–197
Customs, 8, 9, 63

D
Debris, 5, 13, 18, 23, 25, 26, 75, 110, 148, 177, 178
Dispute, 6, 8, 46, 76, 78, 129, 132

E
European Space Agency (ESA), 30, 59, 142, 143
Executive Orders (EO), 58, 64, 70, 73, 74, 103, 106, 116, 129

G
Global South, 40–42, 94–96, 176, 178, 185, 186, 196
Guidelines for the Long-term Sustainability of Outer Space Activities, 4

H
Hague International Space Resources Governance Working Group, 10, 117
Harmful interference, 4, 12, 25, 26, 32, 33, 55, 81, 108, 111, 117, 118, 132, 163, 168
Helium-3, 35, 54, 72, 100, 158

I
Indian Space Policy, 2023 (ISP), 40, 43, 45, 47
Indian Space Research Organisation (ISRO), 93, 94, 164, 165
In-situ, 27, 34, 100, 103, 128, 160
International Court of Justice (ICJ), 7, 9, 46–48, 63, 76–81
International Law Commission (ILC), 6–8, 76, 79, 80, 128
International Lunar Research Station (ILRS), 9, 19, 70, 71, 93, 107, 110, 129, 170
International Seabed Authority (ISA), 44, 45, 48, 60, 65, 133

S. Bhat B., A. Variath (eds.), *Artemis Accords and Resource Mining in Outer Space*, Issues in Space, https://doi.org/10.1007/978-981-95-0152-6

International Space Station (ISS), 9, 108, 109, 143
iSpace, 54, 59

J

Japan's Space Resources Act 2021, 59
Jus cogens, 14

L

Legal Subcommittee (LSC), 63, 73, 184, 186–191, 193–196
Lunar base, 28, 70, 129, 170
Luxembourg's Space Resources Law of 2017, 29, 58

M

Mars, 2, 18, 27, 28, 41, 72, 88, 90, 93, 100, 102, 103, 106, 143, 162, 166–168
Multilateralism, 71, 74, 81, 82, 92–93, 110, 163

N

NATO, 104
Non-appropriation, 11, 40, 43, 47, 64, 90, 92, 124, 125, 128, 131, 132, 159, 160, 163, 166, 168, 183
Non-appropriation principle, 12, 13, 21, 22, 28, 31, 32, 35, 40, 49, 60, 72, 73, 109, 123, 127, 128, 131, 132, 160, 162
Non-governmental entity, 11, 23, 29, 40, 57, 59, 94, 126, 130, 132, 160

O

Opinio juris, 8, 9, 63, 79, 126, 129

P

Pacta sunt servanda, 10, 123
Province of all mankind, 40–49, 53, 56, 80, 89, 123, 138–140, 160, 162, 185
Public-private partnership, 149

R

Res communis, 20, 28, 29, 55, 56, 121, 124–126, 190
Responsible behaviour, 102, 104

S

Scientific data, 3, 18, 23, 25, 55, 74, 79, 146, 150
Space weapons, 102–109
Spacefaring nations, 101, 106, 110, 117, 128, 144, 165, 167, 169, 197
SpaceX, 30, 167, 168
State practice, 6–9, 48, 64, 79, 106

T

Terra nullius, 20
Third World Approaches to International Law (TWAIL), 176–197

U

UN Charter, 77
UNCOPUOS, 4, 106, 110, 141, 182
UN General Assembly (UNGA), 104, 105, 107, 110, 126, 140, 181
United Arab Emirates, 2, 59, 101, 129, 161
United Nations Convention on the Law of the Sea (UNCLOS), 41, 43–45, 48, 60, 74, 81, 89, 90, 125, 196
United Nations Guidelines for the Long-term Sustainability of Outer Space Activities, 2019, 4
United States Space Force (USSF), 104, 164
United States Space Programme, 161–164
US Policy in Outer Space, 105

V

Vienna Convention on the Law of Treaties (VCLT), 5–7, 9, 10, 46–49, 57, 60, 63, 64, 70–82, 123, 124, 127, 128, 160

MIX
Papier aus verantwortungsvollen Quellen
Paper from responsible sources
FSC® C105338

If you have any concerns about our products,
you can contact us on
ProductSafety@springernature.com

In case Publisher is established outside the EU,
the EU authorized representative is:
Springer Nature Customer Service Center GmbH
Europaplatz 3, 69115 Heidelberg, Germany

Printed by Libri Plureos GmbH
in Hamburg, Germany